DATE DUE			
Sep21'76			

ION-SELECTIVE MICROELECTRODES

ADVANCES IN EXPERIMENTAL MEDICINE AND BIOLOGY

Recent Volumes in this Series

Volume 40
METAL IONS IN BIOLOGICAL SYSTEMS: Studies of Some Biochemical and Environmental Problems
Edited by Sanat K. Dahr • 1973

Volume 41A
PURINE METABOLISM IN MAN: Enzymes and Metabolic Pathways
Edited by O. Sperling, A. De Vries, and J. B. Wyngaarden • 1974

Volume 41B
PURINE METABOLISM IN MAN: Biochemistry and Pharmacology of Uric Acid Metabolism
Edited by O. Sperling, A. De Vries, and J. B. Wyngaarden • 1974

Volume 42
IMMOBILIZED BIOCHEMICALS AND AFFINITY CHROMATOGRAPHY
Edited by R. B. Dunlap • 1974

Volume 43
ARTERIAL MESENCHYME AND ARTERIOSCLEROSIS
Edited by William D. Wagner and Thomas B. Clarkson • 1974

Volume 44
CONTROL OF GENE EXPRESSION
Edited by Alexander Kohn and Adam Shatkay • 1974

Volume 45
THE IMMUNOGLOBULIN A SYSTEM
Edited by Jiri Mestecky and Alexander R. Lawton • 1974

Volume 46
PARENTERAL NUTRITION IN INFANCY AND CHILDHOOD
Edited by Hans Henning Bode and Joseph B. Warshaw • 1974

Volume 47
CONTROLLED RELEASE OF BIOLOGICALLY ACTIVE AGENTS
Edited by A. C. Tanquary and R. E. Lacey • 1974

Volume 48
PROTEIN—METAL INTERACTIONS
Edited by Mendel Friedman • 1974

Volume 49
NUTRITION AND MALNUTRITION: Identification and Measurement
Edited by Alexander F. Roche and Frank Falkner • 1974

Volume 50
ION-SELECTIVE MICROELECTRODES
Edited by Herbert J. Berman and Normand C. Hebert • 1974

ION-SELECTIVE MICROELECTRODES

Edited by

Herbert J. Berman
Boston University
Boston, Massachusetts

and

Normand C. Hebert
Microelectrodes, Inc.
Londonderry, New Hampshire

PLENUM PRESS • NEW YORK AND LONDON

Library of Congress Cataloging in Publication Data

Workshop on Ion-Selective Microelectrodes, Boston University, 1973.
Ion-selective microelectrodes.

(Advances in experimental medicine and biology, v. 50)
Includes bibliographical references.
1. Electrodes, Ion selective—Congresses. I. Berman, Herbert J., ed. II. Hébert, Normand C., ed. III. Title. IV. Series. [DNLM: 1. Biomedical engineering—Congresses. W1AD559 v. 50 1973 / TK7874 W926i 1973]
QD571.W7 1973 541'.3724 74-14914
ISBN 0-306-39050-7

Proceedings of a Workshop on Ion-Selective Microelectrodes,
held June 4-5, 1973, at Boston University.

A Division of Plenum Publishing Corporation
227 West 17th Street, New York, N.Y. 10011

United Kingdom edition published by Plenum Press, London
A Division of Plenum Publishing Company, Ltd.
4a Lower John Street, London W1R 3PD, England

Printed in the United States of America

Preface

pH and ion-selective microelectrodes are rapidly finding an increasing number of applications in the study and control of living (and nonliving) systems. They are unique in their capacity to measure chemical species without altering natural or controlled environmental conditions. Furthermore, these potentiometric tools measure the activity of the chemical species in contrast to conventional ones that measure total concentration.

The "Workshop on Ion-Selective Microelectrodes" is designed to provide an insight into the principles, theory, fabrication, techniques, present limitations, goals, and applications of some of these tools.

The importance and types of microelectrodes and guidelines for their application in biological systems are discussed by Berman. Their present limitations are reviewed by Durst. He warns that their use in analyzing living matter should be approached with caution because of the ill-defined nature of biologic systems. Techniques are presented next for the fabrication of pH (Hebert), antimony (Green and Giebisch, and Malnic et al.), oxygen (Whalen), then single-barrelled (Wright, Walker and Ladle, Morris and Krnjevic) and double barrelled (Zeuthen et al., and Khuri) potassium and chloride liquid ion-exchanger microelectrodes. Difficulties with and fabrication of reference and glucose electrodes are covered, respectively, by Durst and Wright, and Bessman and Schultz.

Applications of pH and ion-selective microelectrodes are described in microanalysis (Wright), measurement of intracellular ion activity and calculation of equilibrium potentials (Brown and Kunze), and then studies of the kidney (Wright, Malnic et al., and Khuri), brain (Zeuthen et al., Morris and Krnjevic), frog heart (Walker and Ladle), and human skeletal muscle (Filler and Das). In addition, actual and potential clinical applications, respectively, of pH (Filler and Das) and glucose (Bessman and Schultz) electrodes are reviewed.

It is the hope of the organizers, participants and sponsors, The National Heart and Lung Institute, The Microcirculatory Society, and The Host Institute, Boston University, that publication of the Workshop will focus attention on the potential of this new and rapidly developing area, and on the requisite precautions for the proper use of ion-selective microelectrodes in the study of complex systems.

The Editors

Contents

I. Theory, Potential and Existing State of Development

Perspective: Ion-Selective Microelectrodes: Their Potential in the Study of Living Matter *In Vivo**

Herbert J. Berman

Department of Biology
Boston University
Boston, Mass. 02215

Work with ion-selective macro- and microelectrodes is rapidly evolving into a new specialty area in biology. The inherent advantages of the approach are basic: it enables one to measure activity of an ionic species directly, as opposed to concentration, and it does this simply, quickly, and in the presence of numerous other components in a complex system, negating the need of first isolating the molecular or ionic species of interest. Interfering species can be compensated for in most cases. Selective microelectrodes have the additional unique potential of making measurements of ionic species possible for the first time *in vivo* under conditions that approach normalcy, and, in addition, in a continuous manner. Net reactions can therefore be followed *in vivo*. In contrast, most biochemical measurements today are based on *in vitro* procedures that are nonphysiologic. Furthermore, they are frequently limited to only one point in time. Undoubtedly piercing a cell with a microelectrode can alter its functional state. But, as has been debated in neuro- and electrophysiology, if the microelectrodes are sufficiently fine and the area entered small with respect to the total cell, then the injury should be negligible and the measured activity should approximate that present in the normal living system.

---

*The Organizers and Participants of the "Workshop on Ion-Selective Microelectrodes" acknowledge with thanks sponsorship by the Microcirculatory Society and Boston University, and support by the National Heart and Lung Institute, Labtron, Microelectrodes, Inc., and the Transidyne General Corporation.

Our purpose at this meeting was to examine critically, by presentation and discussion of a limited number of selected review and research papers and by practical laboratory demonstrations, where we stand today in the continuum of development and application of these microsensors and to note where we hope to go.

Types of Electrodes.

There are two basic types of ion-selective electrodes; those composed of a solid barrier across which the potential develops, and those with a liquid barrier, such as a hydrophobic solvent between two hydrophilic solutions. They may be categorized further as shown in Table I.

TABLE I

ELECTRODES

YEAR	TYPE	EXAMPLES OF SPECIES SENSED
	SOLID STATE	
	GLASS	H^+, Na^+, K^+
1906	pH	H^+
1957	OTHERS	Na^+, K^+
1966	CRYSTALS (DOPED)	Cl^-, F^-, I^-, CN^-, SCN^-
1950, 1964	HETEROGENEOUS	Ag^+, Pb^{++}, Cl^-, I^-, S^-, SH^-
	PRESSED PELLET	
	IMPREGNATED POLYMER (PVC, PE, SILICONE RUBBER)	
	GRAPHITE	
	LIQUID STATE	
1966	LIQUID ION EXCHANGE	Ca^{++}, Cl^-, NO_3^-, K^+
1967	NEUTRAL CARRIER	K^+ (VALINOMYCIN, GRAMICIDIN)
		Na^+ (MONENSIN)
	VARIANTS	
	GAS	CO_2, SO_2, NH_4
	IMMOBILIZED BIOLOGICAL REACTANTS (ENZYMES)	GLUCOSE OXIDASE
		B-GLUCOSIDASE
		UREASE
		L- AND D-AMINO ACID OXIDASES

The pH electrode, discovered in the first decade of the century (Cremer, 1906; Haber and Klemenslewicz, 1909), typifies the selective barrier that gives rise to a potential. It was not until 1957 (Eisenman et al., 1957) that glass sufficiently selective to Na^+, and still later to K^+ and other univalent ionic species (Eisenman, 1962) was developed, and fabrication at least of Na^+ glass microelectrodes was considered worthwhile (Lavallée et al., 1969). Earlier, Wyllie and Patnode (1950) discovered a generalized method of making solid ion-selective electrodes. They embedded the selective material in an inert solid membrane to form a heterogeneous membrane barrier. However, because of technical difficulties, progress in this area did not commence until about 1966-1967 (Eisenman, 1969; Ross, 1969). Then heterogeneous semi-microelectrodes were made for many ionic species by compression of the inert material containing different active exchange substances into pressed pellets or polycrystals. Different selective substances were also embedded in a number of inert membrane-adaptable materials, such as silicone rubber, polyvinyl chloride and polyethylene, and then in graphite. The types of electrodes were extended to include doped crystals, such as lanthanum fluoride (Frant and Ross, 1966). In 1967 a most significant advance was made: an inert hydrophobic liquid was substituted for the inert solid material composing the barrier and the ion-selective material incorporated into the matrix of the hydrophobic liquid (Ross, 1967). At the same time selectivity by macrotetrolides (macrocyclic antibiotics) for K^+ and Na^+ was pioneered by Mueller and Rudin (1967), Stefanac and Simon (1967), and others. The dates of the references attest to the relative newness of this specialty area, one in which many ingenious innovations may be expected in the future, most potentially adaptable to the microelectrode level. These include electrodes selective for different polutants in air (Kneebone and Freiser, 1973) and others made selective by immobilization of biological reactants, such as enzymes, within membranes or on their surfaces (Baum and Ward, 1971; Buck, 1972; Mohan and Rechnitz, 1972; Gough and Andrade, 1973; Llenado and Rechnitz, 1973).

Techniques and Criteria for Quantification:

The types of measurements that can be made with the ion-selective probes (Table II) include practically all those presently made today by more routine biochemical procedures. Furthermore, the measurements can be made simply, quickly, continuously with time, and in the case of microelectrodes, in an *in vivo* setting under conditions that are beginning to approach normalcy.

Different techniques, such as those noted in Table III, have or are being developed to obtain data. For example, if one can release bound substances that then ionize, or if one can closely

TABLE II

TYPES OF MEASUREMENT

ACTIVITIES
RATES
GRADIENTS
TRANSIENTS
EQUILIBRIA
(CONCENTRATION)

TABLE III

QUANTITATIVE TECHNIQUES FOR MEASUREMENT

DIRECT
KNOWN ADDITION
KNOWN SUBSTRACTION
RELEASE
DILUTION

approximate the distribution of bound to unbound species by addition of a known amount of an ionic species and estimate the amount bound, then one can approximate the total concentration of the ionic species in a continuous manner.

General criteria, such as those listed in Table IV, should be applied to maximize the reliability of the data obtained with microelectrodes. Questions that should be continuously asked and checked include: the stability of the system. Does it drift? If so, how much? How rapidly? Is the drift random or predictable? How stable is the reference electrode? Other questions relate to the limitations of the system. How accurate is the measurement? What is the minimum activity the system can sense? And discriminate? What is the effect of sample volume on the measurement? Is the reactive surface of the microelectrode totally immersed in the substance being studied? What substances are present that may interfere with the true reading? In what way and to what extent do they alter the true reading? Also, can the calibration fluid be made more representative of the experimental fluid and thereby negate the need for or reduce the magnitude of the corrective factor ($k\ A^{n/z}$ Int) :

$$[E_{cell} = E_o + RT/nF \ \ln(A_{ind.} + k\ A^{n/z}\ \text{Int})$$

TABLE IV

PHYSICAL AND CHEMICAL LIMITATIONS OF SYSTEM OR TECHNIQUE

Proper Functioning of Electrode

Stability of System (Drift, Reference Electrode, Junction Potential, Error)

Selectivity (Interfering Species)

Limits of Sensitivity

Reproducibility (Precision)

Accuracy

Response Time (Rate of Reaction and Degree of Completeness)

Temperature

pH (Buffer) and Total Ionic Strength

where E_o is the constant standard potential for the indicator electrode; R, T, F, and $A_{ind.}$, are the gas constant, absolute temperature, the Faraday constant, and activity of the ionic species being measured; k is the selectivity coefficient; and A_{Int} is the activity of the interfering ions of charge Z. The cell potential thus varies directly with the logarithm of the ionic activity. At 25°C, 2.3 RT/nF has a response slope of 59.2/n mV/pI. That is, at 25°C, the potential of an ion-selective electrode will under ideal conditions change 59.2 mV when the activity of a univalent ion (n) is changed by a factor of ten. (Nernst, 1889; Durst, 1971)]. Assiduous application of these criteria should help to ensure the reliability of the measurement, maximize its accuracy and sensitivity, and hopefully, eventually allow us to achieve the elusive objective of making absolute measurements with high reliability.

Some questions that a physiologist would ask are listed in Table V. He would want to know: Whether or not the system being used can give him the information being sought? How normal the cell or tissue preparation is? And what reliability and confidence he can place on the data being gathered?

Objectives:

Our ultimate objectives are the obvious ones: to determine

TABLE V

PHYSIOLOGIC LIMITATIONS

What are the limitations of the systems and techniques being used? Can they give you the information you seek?

How physiologic is the preparation or cell unit being studied?

Injury - from microelectrode - degree and effect on results?

Exactly where is the microsensing tip located?

What are the physiologic changes with time?
(And those due to the sensor and exposure of the preparation?)

Reproducibility?

Accuracy?

the chemical composition of the material being studied and how it functions without injuring or altering it; also, as an extension of these major objectives, to monitor its functions and learn how to control them so as to optimize the function and the longevity of the system. Furthermore, in applications in pathologic states, we wish to detect the deviation, preferentially in its preclinical state, but, if this is not at first possible, as it occurs, and then to institute corrective measures and monitor their effectiveness, preferably on a continuous and instantaneous basis.

It is unlikely that any one approach will give us all the information needed or sought. Ion selective microelectrodes are a relatively new and rapidly evolving tool applicable to *in vivo* research. It has definite advantages, but like all approaches it has its limitations. For instance, one of the great limitations of several of the present ion-selective electrodes is the concentration range of their response. For example, Ca^{2+} can presently be measured at concentrations as low as $10^{-4.5}$ or 10^{-5}M with reasonable reliability. However, in many cases, we need to sense Ca^{2+} activity rapidly and continuously at concentrations as low as 10^{-8}M. Other new methods applicable to *in vivo* work, some based, e.g., on optics and measurements of transmittance or reflectance at one or more wavelengths, such as in *in vivo* microspectrophotometry (Davila, et al, 1973; Chance, et al, 1971; Liebman, 1969), are also in an early, formative stage of development. It seems most logical that in order to gain greater insight into the complex phenomena challenging us in biology, it would be most advantageous to use all available approaches to elucidate the details of the composition

TABLE VI

OBJECTIVES

THE CHEMICAL AND PHYSIOLOGIC ANALYSIS OF LIVING MATTER AS IT IS.

CONTINUOUS MONITORING OF THE STATE OF BEING AND MECHANISMS OF CONTROL.

MASTERY OF THE CONTROL MECHANISMS.

MAINTENANCE OF THE TOTAL SYSTEM IN ITS OPTIMAL STATE.

and organization of the structures, the details of the reactions within the living system, the precise location at all times of the reactions, the mechanisms of their control, the responses of the reactions to different types and intensities of stresses, and how we may monitor and regulate the reactions *in vivo* at all levels of integration and complexity.

The state of development of our present tools may seem inadequate to achieve our objectives. But it is first by their primitive innovative development, then their use and step-by-step improvement that we have come this far. It is by further refinement of existing approaches, development of new ones, and their ingenious application that one may hope to approach and achieve our ultimate objectives of studying living matter or other materials as they are without production of injury or change, and, in addition, learn how to regulate and optimize their state of being.

REFERENCES

Baum, G. and Ward, F.B., 1971. General enzyme studies with a substrate-selective electrode: characterization of cholinesterase. Anal. Biochem. 42: 487-493.

Buck, R.P., 1972. Ion-Selective Electrodes, Potentiometry, and Potentiometric Titrations. Anal. Chem. Reviews. 44:270R-285R

Chance, B., Lee, C.P. and Blasie, K.J., Eds., *Probes of structure and function of macromolecules and membranes*. Vol. I. Probes and Membrane Function. (Proc. 5th Colloquium of the Johnson Research Foundation. Apr. 19-21, 1969). Academic Press, N.Y. and London, 1971.

Cremer, M., 1906. Uber dreursache der elecktromotorischen Eigenschafter der Gewebe, Zugleichein Beitrag zur Lehre von den polyphasischen elektrolytketten. Z.F. Biol. 47:562-608.

Davila, H.V., Salzberg, B.M., and Cohen, L.B., 1973. A large change in axon fluorescence that provides a promising method for measuring membrane potential. Nature, N.B. 241:159-160.

Durst, R.A., 1971. Ion-selective electrodes in science, medicine, and technology. American Scientist 59:353-361.

Eisenman, G., Rudin, D.O., and Casby, J.U., 1957. Glass electrode for measuring sodium ion. Science 126:831-834.

Eisenman, G., 1969. "Theory of Membrane Electrode Potentials". Chapter 1 in Ion-Selective Electrodes. R. A. Durst, ed., NBS Spec. Publ. 314, U.S. Gov. Printing Office, Washington, D.C.

Eisenman, G., 1962. Cation selective glass electrodes and their mode of operation. Biophys. J. 2 (Part 2): 259-323.

Frant, M.S. and Ross, J.W., 1966. Electrode for sensing fluoride ion activity in solution. Science 154: 1553.

Frant, M.S. and Ross., 1970. Potassium ion specific electrode with high selectivity for potassium over sodium. Science 167:987-988.

Gough, D.A. and Andrade, J.D., 1973. Enzyme electrodes. Science 180: 380-384.

Haber, F. and Klemenslewicz., 1909. Uber elektrische phasengrenzkrafte. Z.F. physik. Chem. (Leipzig) 67:385-431.

Kneebone, B.N. and Freiser, H., 1973. Determination of nitrogen oxides in ambient air using a coated-wire nitrate selective electrode. Anal. Chem. 45:449-452.

Lavallée, M., Schanne, O.F. and Hebert, N.C. 1969. Glass Microelectrodes. John Widey and Sons, Inc.

Liebman, P.A., 1969. Microspectrophotometry of retinal cells. In Data Extraction and Processing of Optical Images in the Medical and Biological Sciences. Annals, N.Y. Acad. Sci. 157:250-264.

Llenado, R.A. and Rechnitz, G.A., 1973. Ion-electrode based autoanalysis system for enzymes. Anal. Chem. 45:826-833.

Mohan, M.S. and Rechnitz, G.A., 1972. Ion-electrode study of the calcium-adenosine triphosphate system. J. Am. Chem. Soc. 94: 1714-1720.

Mueller, P., and Rudin, D.O., 1967. Development of K^+-Na^+ discrimination in experimental bimolecular lipid membranes by macrocyclic antibiotics. Biophys. Biochem. Res. Comm. 26:398-404.

Nernst, Walther, 1889. Die elektromotorische Wirksamkeit der Ionen. Z.F. physik Chem. (Leipzig) 4: 129-181.

Rechnitz, G.A. and Mohan, M.S., 1970. Potassium-adenosine triphosphate complex: formation constant measured with ion-selective electrodes. Science 168:1460.

Ross, J.W., 1969. "Solid-state and liquid membrane ion-selective electrodes". Chapter 2 in Ion-Selective Electrodes, R.A. Durst, ed., NBS Spec. Publ. 314, U.S. Gov. Printing Office, Washington, D.C.

Ross, J.W., Jr. 1967. Calcium-selective electrode with liquid ion exchanger. Science 156:1378-1379.

Stefanac, Z. and Simon, W., 1967. Ion specific electrochemical behavior of macrotetrolides in membranes. Microchem. J. 12:125-132.

Wyllie, M.R.J. and Patnode, H.W., 1950. Equations for the multiionic potential. J. Phys. Chem. 54:204-227.

ION-SELECTIVE ELECTRODE RESPONSE IN BIOLOGIC FLUIDS

Richard A. Durst

Analytical Chemistry Division
National Bureau of Standards
Washington, D.C. 20234

ABSTRACT

The response of ion-selective electrodes in biological fluids may be affected by a wide variety of physical and chemical factors. These may influence the indicator electrode directly or may affect the liquid junction of the reference electrode. A brief discussion is presented of the various sources of error and uncertainty in electrode measurements in biologic media especially with microelectrodes. A serious need exists for the development of practical standards for the calibration of ion-selective electrodes in physiologic media in order to ensure the consistency of interlaboratory measurements.

I. INTRODUCTION

While the behavior of ion-selective electrodes in simple aqueous solutions is fairly well understood, there are still many uncertainties in the interpretation of electrode measurements in biologic fluids, especially microelectrode measurements in vivo. Ideally, the emf response of ion-selective electrodes is directly proportional to the logarithm of the activity of the ionic species, as given by the Nernst equation:

$$E_{ISE} = \varepsilon + \frac{RT}{zF} \ell n\ A$$

or more rigorously by the Nicolsky-type equation:

$$E_{ISE} = \varepsilon + \frac{RT}{zF} \ell n\ [A_i + \sum_{j=1}^{n} k_{ij} A_j^{z/y}],$$

where E_{ISE} is the measured emf of the ion-selective electrode, ε is a constant term which incorporates the standard potential characteristics of the electrode, RT/zF is the Nernstian response factor (R, gas constant; T, absolute temperature; F, the Faraday constant), z is the charge on the primary ion including its sign, y is the charge on the interfering ion, A_i and A_j are the activities of the primary and interfering ions respectively, and k_{ij} is the selectivity coefficient.

In the study of many biologic phenomena, it is the activity of a given ionic species which is desired, but the measurement of this quantity by ion-selective electrodes is fraught with many uncertainties, both conceptual and experimental. Without going into great detail, I should like to give an overview, as seen by a non-specialist, of some of the more serious difficulties encountered in biological studies with ion-selective microelectrodes. In general, the precision of microelectrode measurements in biologic media is quite good, whereas accuracy is a much more complex parameter which requires considerable ingenuity on the part of the experimenter to avoid a multitude of potential sources or error. For a more detailed discussion of the principals and techniques involved in electrode measurements of this type, several excellent monographs are available and highly recommended [1].

II. INDICATOR ELECTRODES

With respect to the ion-selective electrodes themselves, there are several sources of error to be considered. First, the ions of interest may be in a form to which the electrode is insensitive. For example, if the ions are in an insoluable or complexed form, such as bound to cellular constitutents or structures (e.g., the cell membrane), or complexed with polyelectrolytes (i.e., proteins), and other ligands (i.e., complexing agents and counter ions), then they will not be sensed by the electrode. In other words, ion-selective electrodes respond only to the free ionic species in solution. Similarly, if the ionic species is not distributed homogeneously throughout the sample, the electrode will respond only to that ionic activity existing at the sensing tip of the electrode. Thus, if heterogeneous cellular ion distribution occurs because of compartmentalization, comparison of microelectrode measurements with total concentration methods will not be valid. The fact that ionic distribution depends on the particular membrane-enclosed compartment into which the microelectrode is inserted (e.g., cytoplasm, extracellular space, nucleus, etc.) necessitates extreme care in trying to evaluate activity coefficients by intercomparison of activities determined by electrodes versus concentrations measured by "averaging" techniques such as flame photometry or atomic absorption.

The response of an ion-selective electrode may also be affected by the presence of proteins and other organic constituents which can coat the surface of the electrode membrane. This very often leads to a sluggish response, but may also shift the observed emf due to the presence of active sites in the coating molecules and changes in membrane selectivity caused by the superimposition of the organic coating characteristics onto the normal electrode behavior. This is a very complex problem which requires considerable research effort to elucidate all of the possible interactions.

A very common error is that caused by ionic species to which the electrode is sensitive. That is, ion-selective electrodes are not specific for one particular ion and may respond to other ions in varying degrees of sensitivity. The pH glass electrode is probably the most specific electrode available in terms of interferences by other ions. Only in highly basic solutions, i.e., extremely low hydrogen ion activity, does the effect of alkali metal ions, such as sodium, become apparent. The newer types of non-glass membrane electrodes are even more sensitive to this type of interference. In most cases, however, manufacturers have determined selectivities for the commercial electrodes over the most common interfering ions, at least to a first approximation. Care must be exercised in using these selectivity coefficients because they are dependent upon experimental conditions, e.g., ionic strength, relative ionic concentrations, etc., and in some cases may actually be reported as the reciprocal of the selectivity coefficient as defined in the Nicolsky equation. While these coefficients cannot be used to correct the observed emf for known concentrations of interfering ions, they are useful in anticipating errors from such interferences and taking remedial action such as masking the interferences and buffering the sample.

Another possible source of error is that due to the presence of "bound" water resulting in anomalous ionic activities when compared to solutions of similar ionic strength in vitro. That is, water which is bound or structured by the presence of various cellular components is no longer "free" and available to act as a solvent. This effect results in anomalous activity coefficients when ionic strength, apparent solvent volume, and ionic concentrations determined by non-electrode techniques are interrelated. The activity of water, or its osmotic coefficient, in cells is an extremely complex and difficult problem to study mainly because of the nature of the sample which is segregated in a variety of microcompartments in the cells.

Finally, changes in ion binding and other cell parameters in disease states can also complicate the interpretation of the data and result in errors. Again, while it is possible to make many microelectrode measurements precisely and then to use deviations from normal values as a means of clinical diagnosis, little is really known at the present time in terms of the absolute accuracy of these measurements.

III. REFERENCE ELECTRODES

An even more serious source of error in biologic fluid measurements by electrodes is the reference electrode. The biological scientist is probably more aware of, and frustrated by, reference electrode problems than any other type of electrode user. While the reference element of the electrode, be it mercury/calomel or silver/silver chloride, is usually stable, well-defined and trouble-free, the liquid junction between the internal reference solution and the sample solution is rather poorly understood except for the simplest electrolyte systems.

At the present state-of-the-art, the problems associated with intracellular reference microelectrodes are virtually insurmountable insofar as accurately defining the absolute value of the liquid junction potential. In terms of precision, the microcapillary reference electrodes are quite good. It is only when one is interested in the "true" or absolute activities of ionic species in biologic fluids that real problems arise. Attempting these types of measurements can best be described as "bootstrap" operations. Even in the most ideal case, that of hydrogen ion activity measurements where the defining hydrogen-gas/platinized-platinum electrode can be used, as soon as one interposes a liquid junction between the indicator and reference electrode half cells, an uncertainty is introduced which increases with increasing dissimilarity between the solutions on either side of the liquid junction.

Normally, the reference electrode half cell is connected to the sample solution by means of a salt bridge. Ideally, the salt bridge should produce a minimum liquid junction potential, preferably zero and stable. The ionic constituents of the salt bridge should not poison or otherwise alter the sample. For example, in whole blood measurements, a hypertonic salt bridge solution may cause crenation of the blood cells and protein coagulation, while a hypotonic salt bridge solution may cause hemolysis of the blood cells. An isotonic salt bridge solution, on the other hand, is less well able to minimize the liquid junction potential changes caused by variations in the sample solution composition and is apparently more sensitive to polyelectrolytes and colloids. Of course, the salt bridge constituents must be ions to which the ion-selective electrode is insensitive. For example, although RbCl has been used as the salt bridge electrolyte for measurements with the potassium ion glass electrode, this salt should not be used with the newer valinomycin-type potassium ion electrodes since this sensor is highly selective for rubidium ions.

While it is not too difficult to find the conditions necessary to satisfy these latter two requirements, minimizing the liquid junction potential can often be a very frustrating experience. Many of the problems associated with the liquid junction are common to all reference electrodes, while others are confined primarily to the microcapillary electrodes and/or measurements in fluids containing polyelectrolytes and colloidal or suspended components. There are several factors which affect the liquid junction potential of most reference electrodes. One of the most important factors is the mobility and concentration of the bridge electrolyte vis-a-vis the sample electrolyte. Usually the salt bridge solution is chosen to have a high concentration of equitransferent ions. In this way, conditions are established for the transport of charge at the liquid junction by the salt bridge electrolyte, which, if equitransferent, results in a minimal diffusion potential.

Another factor which may influence the liquid junction potential is termed the "suspension effect" in which the presence of colloids or suspended particles, e.g., red blood cells, produce an anomalous liquid junction potential. It has been suggested that this phenomenon is caused by the effect of colloidal particles on the relative rates of diffusion, i.e., transference numbers, of the salt bridge electrolyte. Another possibility is that colloids with ion-exchange properties give rise to a Donnan potential across the suspension/supernatant liquid interface. Whatever the cause, the effect may be significant and must be avoided in accurate studies with electrodes.

Streaming potentials are another source of uncertainty which are apparently attributable to changes in the liquid junction caused by variations in the flow rate of the sample solution. Factors which influence the observed emf variations include the magnitude of the flow rate changes and the geometry of the electrode system. Although it is known that streaming occurs in the cytoplasm of cells, it is unlikely that this effect is of sufficient magnitude to be a serious problem with intracellular measurements with microelectrodes.

The tip potential of a capillary microelectrode is a very serious source of uncertainty in the determination of intracellular ion activities and in the accurate measurement of membrane and action potentials. Studies have demonstrated that the magnitude of the tip potential depends on the tip diameter, the type of glass used to make the capillary, and the composition and concentration of the solution in the capillary and in the sample. While the absolute value of the tip potential cannot be ascertained at present, it can be minimized by working with a capillary having the largest tip diameter feasible and by calibrating the electrode in a solution as similar as possible to the sample solution. As a further check, the capillary microelectrode should always be recalibrated after the completion of measurements to ensure that the tip was not plugged or broken during penetration of the cell wall.

One possible way to avoid some of the problems described above would be to use an electrode pair without a liquid junction, i.e., a cell without transference. In this way, uncertainties due to the liquid junction, such as alteration of the sample solution by electrolyte diffusion, streaming potentials, suspension effect, and the liquid junction potential itself, may be eliminated by using a pH or other ion-selective electrode as the reference electrode. The difficulty in this approach arises because, in order to assign an accurate emf value to the reference electrode, the activity of the reference ion in the sample solution must be accurately known and remain constant. Once again we are confronted by the necessity of a bootstrap operation. There is no way, at the present state-of-the-art, to accurately calculate the activity of an ion in such a complex mixture as a biologic fluid. If an activity is arbitrarily assigned to the reference ion and if it remains constant, then such an electrode system can be used for precise measurements of relative ion activities, but little can be said about the absolute activities.

Another technique that may be used to get a handle on the activity coefficient is the microinjection of a known concentration of salt containing the ion of interest. If one knows the sample volume and the activity of water in the sample compartment, then the activity coefficient can be calculated from the observed emf change. Unfortunately, these two conditions are rarely known or easily determined for cellular samples, and the injections of concentrations sufficient for reliable measurement probably perturbate the cell sufficiently that correlations would be questionable.

Another difficulty often encountered in many types of electrode measurements is drift of the electrode potential. This problem is usually attributable to the internal reference electrode or the liquid junction. Drift can usually be categorized as one of three types. Parallel drift is characterized by a shift of the calibration line for an electrode system in which the slope of the line does not change. If this type of drift is small enough and unidirectional, it is often possible to make corrections to the sample emf readings by recalibrating the electrode system often and applying a time-dependent correction to the observed potentials. A concentration-dependent drift occurs when the slope of the calibation line changes. For example, one might observe a negative emf drift at high concentration and a positive drift at low concentration. This type of drift in effect results in a pivoting of the calibration line around some intermediate point, but is seldom regular enough to allow a correction to be made. Normally, the drift results in a decrease in the slope of the calibration line indicating a degradation in the electrode response (sub-Nernstian). This effect is most commonly observed with liquid ion-exchange type electrodes and necessitates renewal of the ion-exchanger filling solution. Random drift is characterized by no regular trend, i.e., nonuniform both in direction and magnitude of drift. It is impossible to correct reliably and requires replacement of the defective electrode(s).

The causes of electrode drift, e.g., unstable liquid junction, temperature gradients, high impedance, non-equilibrium system, electrode malfunction, etc., are numerous and varied. In general, replacement of one or both of the electrodes will remedy the problem, otherwise, sample instability or interfering sample constituents should be considered as a possible source of the difficulty.

IV. CONCLUSIONS

Obviously, much more work is needed in this area before ionic activities can be measured accurately. Serious consideration must be given to interionic, ion-solvent, and ion-protein interactions in order to resolve the differences observed for activity coefficients in various biologic media.

One possible direction that should be followed to improve the accuracy of electrode measurements in biologic fluids is the development of physiologic standards which match, as closely as possible, the biologic samples of interest. As with the operational pH scale, the majority of practical acidity measurements need only be reproducible and consistent from one laboratory to another. If suitable physiologic standards can be developed, the operation definition of the pIon scale would be:

$$\text{pIon(X)} = \text{pIon(S)} + \frac{E_X - E_S}{(RT\ln 10)/F}$$

where the pIon of an unknown solution X is calculated from that of an accepted standard (S) and the measured difference in emf (E) for the cell

$$\text{RE} \,||\, \text{soln. } {}^{S}_{\text{or } X} \,|\, \text{ISE}$$

when the standard solution is removed from the cell and replaced by the unknown solution. This definition is strictly valid only when the liquid junction potential between the reference electrode and unknown solution exactly matches that of the standard solution. Routine correction for the residual liquid junction potential is not generally practical. From the foregoing discussion, it is clear that the development of such standards will be no simple task and will require the interaction of biologists, physiologists, solution thermodynamicists, analytical and electrochemists. A tremendous effort will be needed to accurately define the ionic activities of mixed electrolyte systems containing a variety of organic constituents. Initially at least, such standards must of necessity be arbitrary to some extent but, by careful choice of systems and conventions, this can be minimized.

The general tone of the above discussion has been rather negative in order to emphasize some of the pitfalls which must be recognized and avoided in measurements with microelectrodes. This does not mean that such measurements are worthless from the point of view of providing useful information and data on biologic fluids, especially intracellular and in vivo measurements with microelectrodes. Indeed, as stated above, the precision of such measurements is quite good, and as long as electrode calibration is carried out in a reasonable and consistent way, considerable information and knowledge can be gained by electrode measurements, e.g., the effects of physical and chemical perturbations, normal vs. abnormal ionic activities, and ionic distributions in cellular compartments and fluids. Nonetheless, it is imperative that more research go into establishing realistic standards which can greatly improve the absolute accuracy of these techniques. In the next few years, as more of the newer ion-selective electrodes are miniaturized for such biologic measurements, it is likely that research in these techniques will answer many of the problems discussed above and these sensors will be used much more routinely for medical research and clinical diagnosis.

REFERENCES

Bates, R. G., Determination of pH, Theory and Practice, 2nd ed., John Wiley & Sons, New York (1973).

Durst, R. A., editor, Ion-Selective Electrodes, U.S. Government Printing Office, Washington, DC (1969).

Eisenman, G., editor, Glass Electrodes for Hydrogen and Other Cations, Marcel Dekker, Inc., New York (1967).

Feder, W., editor, Bioelectrodes, Ann. N. Y. Acad. Sci. 148, 287 pp. (Feb. 1, 1968).

Geddes, L. A., Electrodes and the Measurement of Bioelectric Events, John Wiley & Sons, New York (1972).

Lavallee, M., Schanne, O.F., and Hebert, N. C., editors, Glass Microelectrodes, John Wiley & Sons, New York (1969).

GLASS MICROELECTRODES FOR pH

Normand C. Hebert

Microelectrodes Inc.

Londonderry, New Hampshire

INTRODUCTION

The important requirements of an intracellular pH microelectrode are:

A. A sharp tip to minimize cell wall damage on penetration.
B. Short pH sensing length so that it will be completely within the cellular compartment being measured.
C. A smooth transition between the pH sensing portion and the insulated portion.
D. A conical angle as small as possible for the portion of the microelectrode that will be in the cell wall. A small angle will make a better microelectrode to cell wall seal thereby preventing leakage of intracellular fluid and shunting of the microelectrode (1).
E. A fast and reproducible response to pH and a design that can be easily cleaned and rejuvenated.

Three types of glass microelectrodes have been used to measure pH. These are the spear, the internal capillary, and the recessed-tip microelectrodes.

The spear-type glass microelectrode (Hinke-type) (2) (Figure 1a) has been used to determine the intracellular pH of the crab muscle fibers (3,4), cytoplasm of the giant squid axon (5), rat sartorius muscle fibers (6), rat atrial muscle fibers (7), vacuole of the algae *Nitella flexilis* (8), rat kidney tubular fluid (9,10), nerve cells of mollusk ganglia (11), and skeletal muscle fibers of the rat (4,12,13,14).

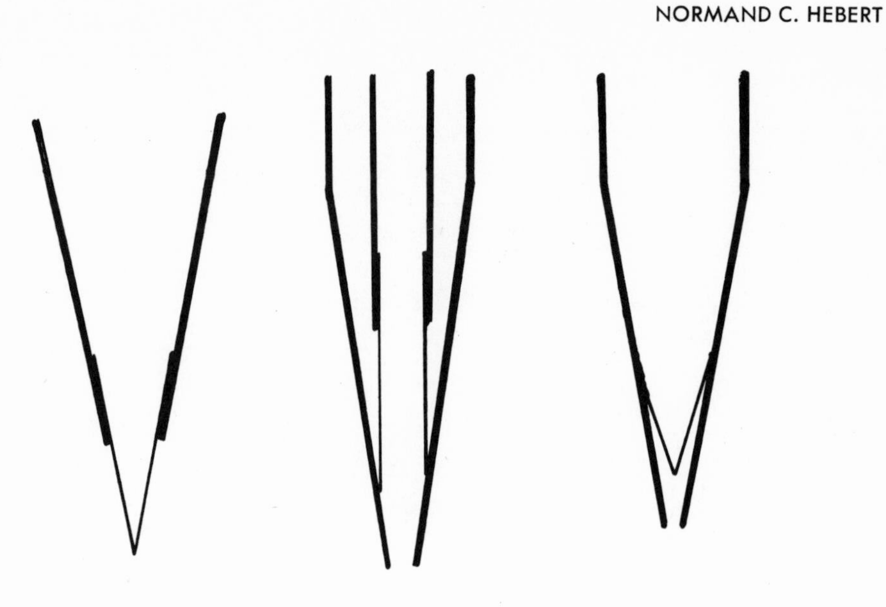

Figure 1. Types of pH glass microelectrodes. In the spear-type microelectrode (1a), the pH sensing tip protrudes from the end of a glass micropipette. In the internal capillary-type pH microelectrode (1b) the pH sensing capillary is sealed inside the shoulder of a glass micropipette. In the recessed-tip microelectrode (1c), a sealed-tip pH sensing pipette is sealed inside the shank of an insulating micropipette. The thin lines indicate the pH sensing glass.

Internal capillary-type glass microelectrodes, (Khuri-type) (15) (Fig. 1b), have been used to determine the intraluminal pH of rat kidney tubular fluid (15), and the hemolymph of the female spider (16).

The recessed-tip glass microelectrodes, (Thomas-type) (17) (Fig. 1c), have been used to measure intracellular pH in perfused liver (18).

The selection of the type of microelectrode to use in a given experiment is best determined by the application, since each type of electrode has advantages as well as limitations.

The sharpest tip can be obtained with the spear-type microelectrode since it is formed with the microforge. Tip diameters as small as one micron or less can be formed. Larger tip diameters are necessary for the internal capillary electrode and the recessed-tip electrode. The internal capillary electrode is designed for suctioning the test fluid into the electrode. The recessed-tip electrode requires equilibration of test fluid with the distilled water which occupies the volume between pH sensing glass and the tip of the outer micropipette. The internal-capillary electrode tip is usually bevelled to 10 microns. The recessed-tip electrode generally has a tip diameter of 2-3 microns.

The shortest sensing length can be obtained with the internal capillary and recessed-tip microelectrodes. With these, it is necessary that the very tip of the electrode is in contact with the test fluid. On the other hand, with the spear-type microelectrode it is necessary to have the complete sensing length inside the test fluid. We have made these electrodes with sensing lengths of about 10 microns. However, more commonly sensing lengths of 50 microns are made.

We find all glass electrodes to be fast-responding as long as the electronics is capable of handling the electrical signal. A spear-type pH microelectrode having a sensing length of about 50 microns has a response time of less than 15 seconds on a Keithly electrometer Model 610B.

Slow responding electrodes are easily rejuvenated and cleaned by dipping in warm (60°C) chromic acid cleaning solution for about two minutes. The electrodes are then exercised by dipping alternatively in hydrochloric acid (0.1 N) for about one minute and sodium hydroxide (0.1 N) for about one minute. The exercising is continued until an optimum response time is observed. The electrode is then stored in the hydrochloric acid solution before standardization.

This technique works well with the spear-type and the internal capillary microelectrode. The response time of the recessed-tip microelectrode is dependent upon the rate of mixing of the distilled water which occupies the space between the pH glass and the outer micropipette tip. The rate of mixing is thus determined by the inside diameter of the outer micropipette tip which is about 1-2 microns.

The easiest electrode to clean is the spear-type. This is done by dipping it in warm chromic acid cleaning solution for about two minutes. The electrode should then be exercised as described above until stable reproducible readings are obtained. The internal capillary microelectrode can be cleaned by suctioning chromic acid into it and rising it out with water followed by exercise. I have not had any experience cleaning recessed-tip microelectrodes, but assume that chromic acid could be used.

We have devoted much time in further developing the method of Hinke (2) for the fabrication of the spear-type pH microelectrode. I feel that this electrode, in spite of its limited applications which is due mainly to its long sensing length, has many advantages over the other two types. The fabrication technique can be further improved to produce shorter sensing lengths. This paper deals with the techniques used in our laboratory for making spear-type pH microelectrodes. Our techniques are based mainly on those developed by Hinke (2) and de Fonbrune (19).

APPARATUS & MATERIALS

The microforge is shown in Figure 2. It sits on a steel plate (18" x 24") which covers a two inch thick marble slab (18" x 24"). The added weight of the marble further isolates the equipment from vibration. The system consists of a Wild M5 microscope with a maximum magnification of 150X and two Narishige MM3 micromanipulators; a left handed one for the microforge and right handed one for the glass tubing; two light sources, one incident and one direct and a microforge. The microforge itself consists of a 25 micron diameter Iridium platinum wire. The wire is placed on a holder (Figure 3) which fits into the left handed micromanipulator. The two wires can be attached to the holder so that they can electrically heat the Iridium platinum wire. Heating is accomplished by means of a variac (General Radio WRMT) and a step-down transformer (Stancor P-6134). Prior to using the holder we simply soldered the small heating wire onto the end of two wires which were taped onto a glass rod.

Glass Tubing:

Two types of glass are necessary for making glass microelectrodes, pH glass and insulating glass.

The pH glass used is Corning Code 0150. This glass is generally available in the form of tubes and has a fairly wide temperature working range. The newer pH glass, i.e. those containing lithium oxide, are not available in the form of tubes and generally have too narrow a temperature working range to be pulled into microelectrodes.

The insulating glass used is Corning Code 0120. It is readily available in tube form and has a thermal expansion co-efficient and softening point, close to that of the pH glass. (Table 1). Ideally the pH glass should have a much higher softening point than the insulating glass. This would make it easier to seal the insulating glass around the pH glass.

TABLE I

Properties of pH and Insulating Glasses

	Thermal Expansion Coefficient	Softening Point
	10^{-7}in./in./°C	°C
Corning Code 0150 (pH)	110	655° (20)
Corning Code 0120	89	630° (21)

The pH glass tubing selected is 1.0 mm O.D. X 0.5 mm I.D. The insulating glass tubing is 2.5 mm O.D. X 1.5 mm I.D. We find that a glass-to-glass seal is more easily made when the outer insulating glass pipette has a thinner wall than the inner pH glass pipette. This compensates for the two glasses having similar softening properties. The softening point of a glass is roughly the temperature at which glass adheres together and begins to flow under its own weight. The definition is actually more exact than this, but this one is sufficient for our purpose. Thus, if one heats two glass fibers having similar softening points, but one is twice the diameter of the other, the smaller fiber will elongate faster than the other. This situation is exactly what we have and is exactly what we do not want. We circumvent this problem to some extent by using a thick wall pH glass capillary and a thin-walled insulating glass capillary.

TECHNIQUE

Fabrication of the Outer Insulating Pipettes

Glass tubing, Corning Code 0120, 2.5 mm O.D. X 1.5 mm I.D. is cut into five inch lengths and cleaned in warm chromic acid solution. The tubing is then washed many times with distilled water and dried in an oven. A large test tube is used as a cleaning container and also for storing the clean tubes.

About twelve micropipettes are pulled on a vertical electrode puller such that pipette lengths from shoulder to tip are about one centimeter with a uniform conical angle near the tip of the pipette. The micropipette tips are then broken off at about 40 microns in diameter. This is accomplished by loading a globule of glass onto the microforge wire such that it surrounds the wire (Figure 4a). The microforge wire (0.005" platinum) is heated to a red color and the micropipette is jammed into the hot globule of glass to the

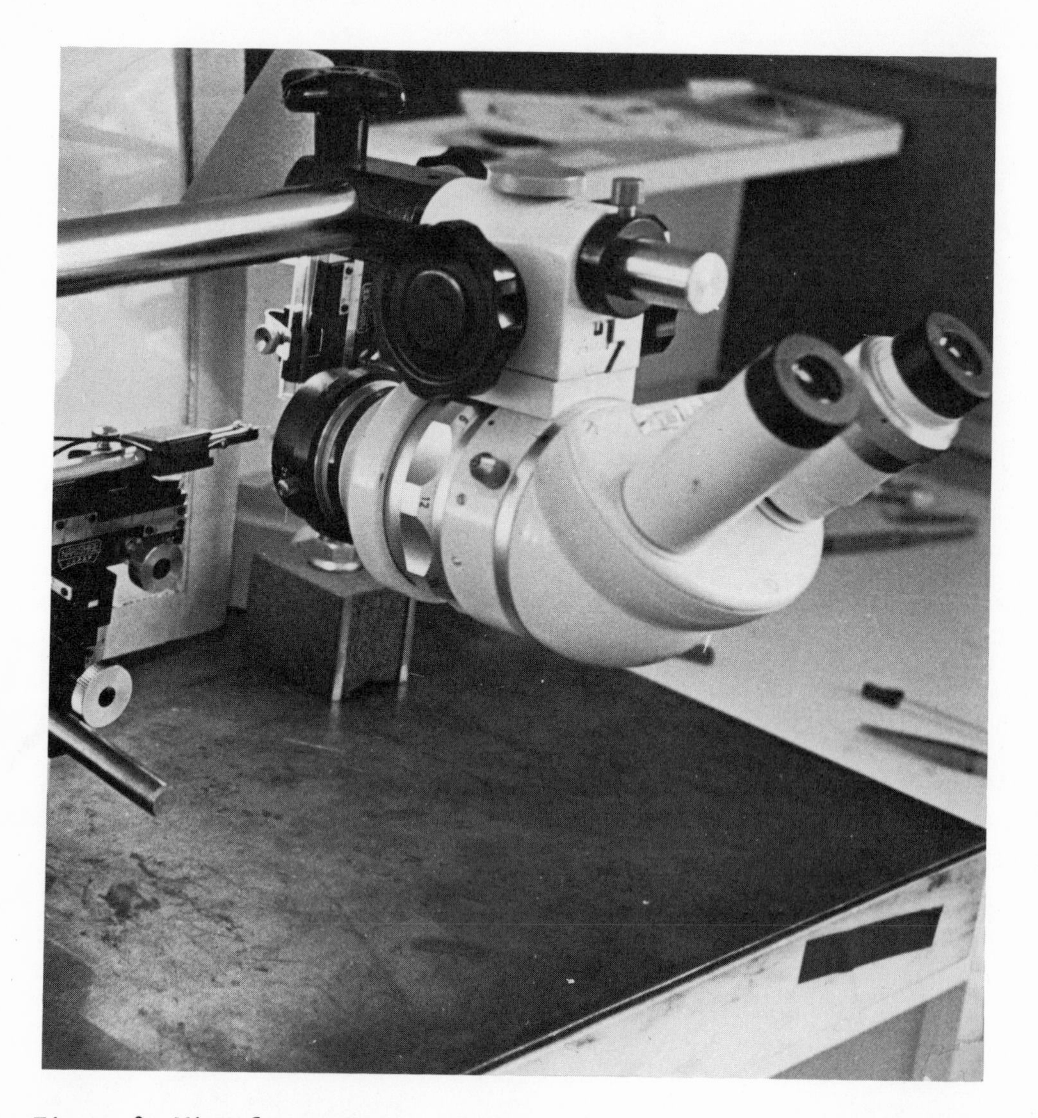

Figure 2. Microforge set-up consists of a Wild M5 microscope with 150x maximum magnification; A source of reflected light and transmitted light; (The latter is located behind the translucent screen). Two Narishige MM3 micromanipulators and a microforge. The equipment rests on a steel plate which sits on a two inch thick marble slab.

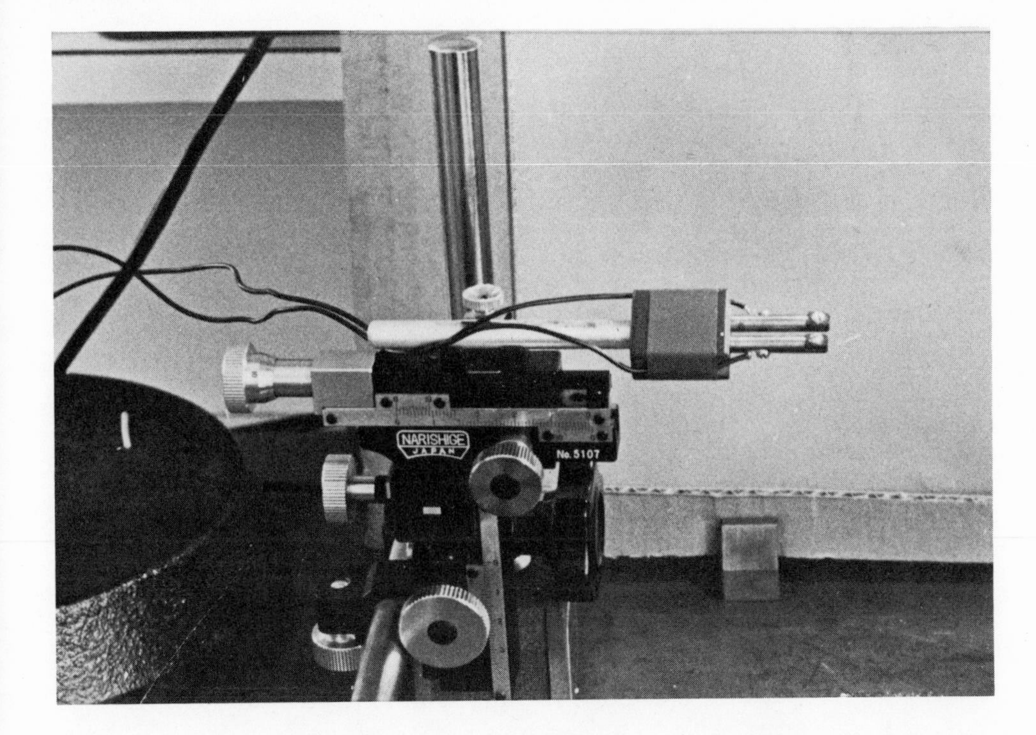

Figure 3. Microforge holder. The Microforge holder was designed by Mr. Placide A. Hebert. It consists of an aluminum rod (3/4" diameter) attached to a plexiglass cube which holds two copper tubes. Two regular screws secure lead wires needed for heating. Two small screws with square washers permit squeezing the heating wire.

point where the pipette is near 40 microns in diameter (Figure 4b). The microforge is then shut off and the micropipette is pulled away from the globule of glass until it breaks off (Figure 4c). The important thing is to have a sharp break, not angular or jagged. The pipette can be broken to a maximum of about 50 - 60 microns diameter using the 0.005" diameter wire. A larger wire has to be used for breaking tips at a larger diameter. Otherwise, the wire is pulled off the forge before the glass breaks.

Fabrication of Inner pH Pipettes

Long pipettes are pulled on the vertical puller using Corning Code 0150 glass tubing (1 mm O.D. x 0.5 mm I.D.). A pH pipette is cut at the shoulder using a sharp scribe. The pipette is then inserted into the outer pipette, and by tapping the outer pipette, it can be made to protrude out of the broken tip of the outer pipette

4a 75X

4b 150X

4c 150X

Figure 4. Breaking of Outer Insulating Pipette Tip

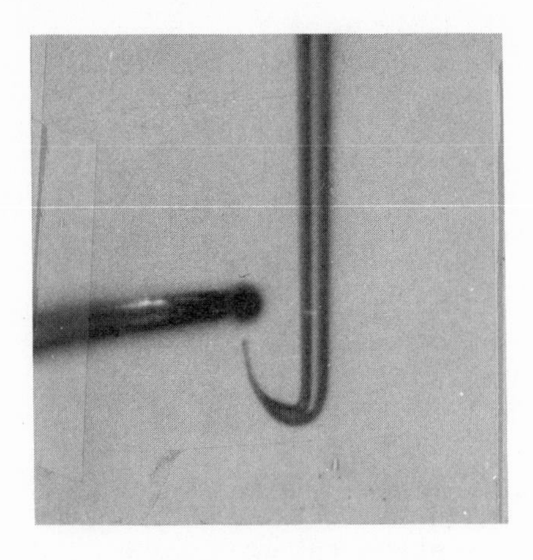

5a

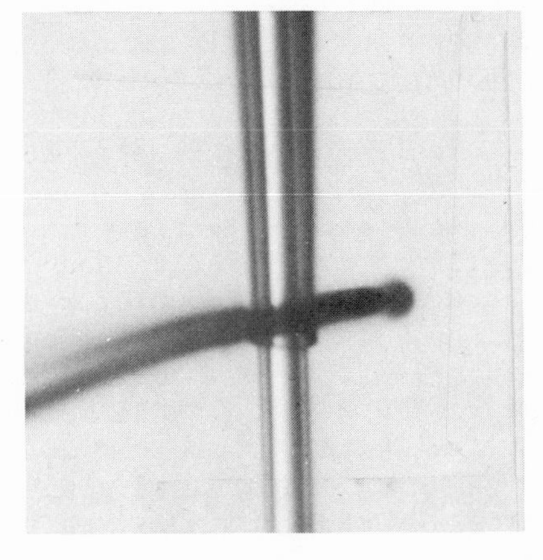

5b

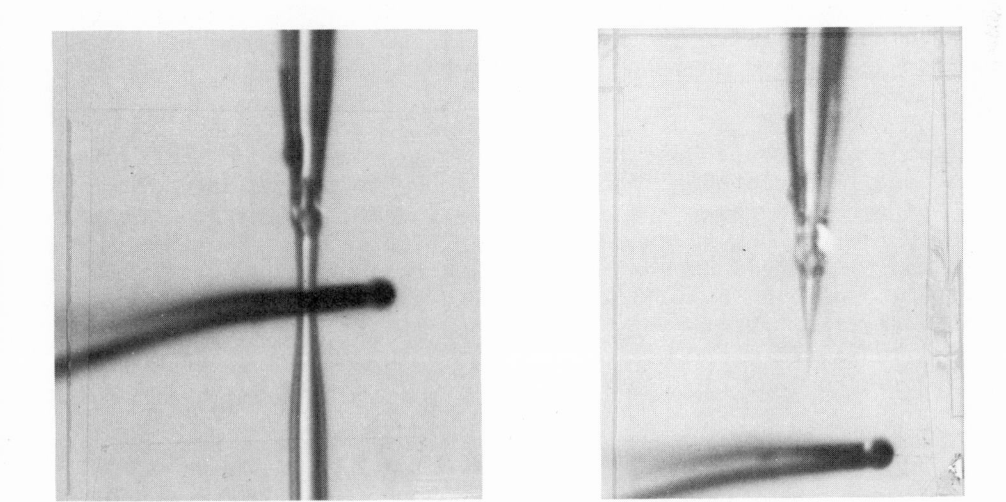

5c

5d

Figure 5. a. A small hook is formed on the end of the pH pipette. (150X) b. The pH pipette is inserted inside the V shaped microforge wire. A small weight (200 mg) is added. The two glasses are sealed together by heating with the microforge wire. (150X) c. The microforge wire is lowered and heated to stretch the pH glass (150X). d. The pH pipette elongates and eventually separates. (150X).

for several millimeters. The important thing is that the inner pH pipette fits tightly at the open-end of the outer pipette. The inner pipette should not hang loosely.

Sealing of pH Pipette to Insulating Outer Pipette

The insulating pipette with the inserted pH pipette is mounted vertically on the right handed micromanipulator. With the use of the microforge wire (25 micron diameter Platinum 90% Iridium 10%) a small hook is formed on the end of the pH pipette (Figure 5a). The micropipette with hook is then inserted inside the V of the microforge wire and a small weight (200 mg) is attached to the hook. The weight is made by sealing a wire loop (25 micron wire) inside a short end of lead glass tubing (Corning Code 0120). The microforge wire is then raised to the point where the inner pipette meets the outer pipette (Figure 5b). Care must be taken to make sure that the pipette is heated evenly all around. The wire is heated to a dull red or to a point where the glass pipette bends slightly. On raising the temperature of the wire slightly, the two glasses will stretch and seal together. Care must be taken not to overheat and cave the glass in at this point. The sealing step should be done slowly. The wire can be moved up and down but not more than 25 microns in each direction. The sealing can take 2 to 3 minutes.

After sealing, the heating wire is lowered about 20 microns below the seal and heated slightly (Figure 5c) to the point where the pH pipette elongates and separates. The pH sensing pipette will generally have a long fiber sensing tip (Figure 5d). The sensing length can be reduced by pushing the electrode horizontally into the hot wire and pulling away from the cooled wire. This is done repeatedly until a satisfactory sensing length and sharp tip are obtained. This will yield microelectrodes with sensing lengths of about 50 microns and tip diameters of 1-3 microns (Figure 6).

The microelectrode tip is then placed into water for about 20 seconds to check the glass seal. It is then examined under the microscope for water which might have leaked in by capillarity. Electrodes which absorb water are discarded. Those that are well sealed are placed on a plexiglass holder for filling and testing.

Filling and Testing of Microelectrodes

The microelectrodes held in the plexiglass holder are placed tip down in a beaker. The beaker is filled with distilled water and the water is boiled for about 20 minutes. The container is allowed to cool in air for about one hour. While the container is still warm, the electrodes with holder are removed from the beaker. A 0.1 N HCl solution is then placed in each electrode using a syringe

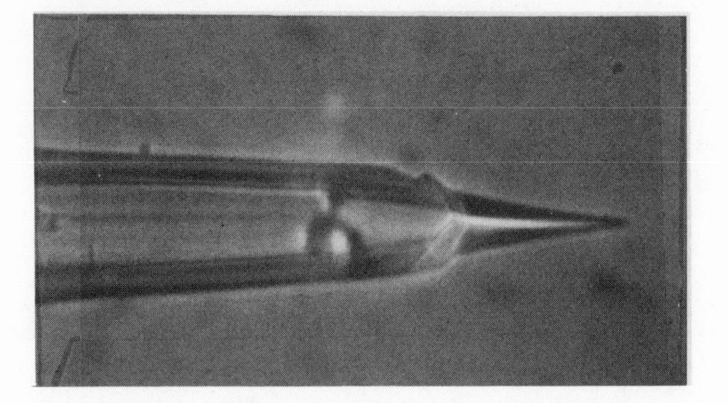

Figure 6. Photomicrograph (500X) showing tip portion of pH microelectrode having a tip diameter of about 2 microns and a pH sensing length of 50 microns.

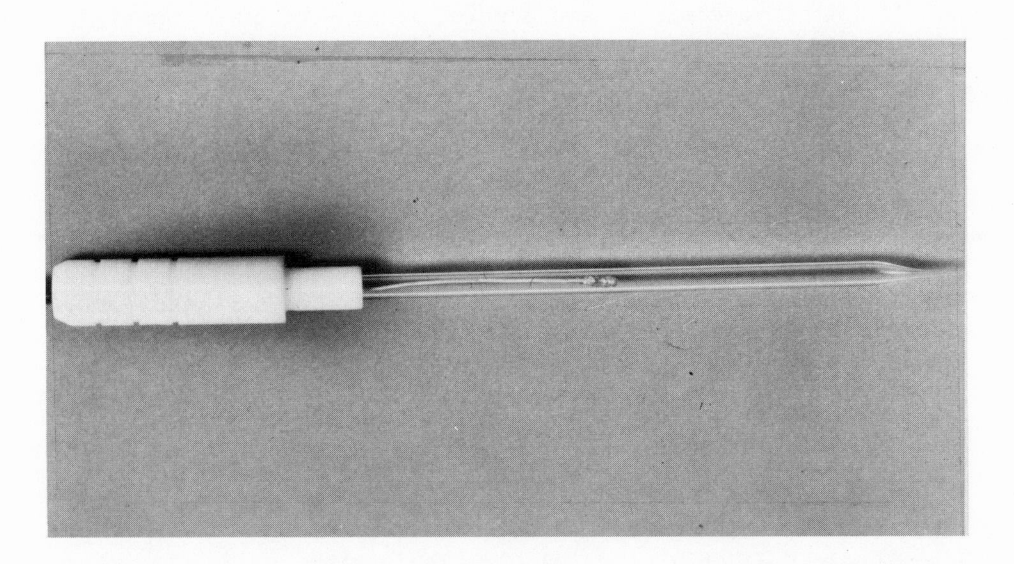

Figure 7. Photograph of microelectrode with silver-silver chloride inner reference electrode, teflon cap, and coaxial cable.

with a long glass needle. This replaces most of the distilled water with hydrochloric acid solution. The electrodes are then allowed to cool to room temperature and tested.

A silver-silver chloride electrode is inserted into the electrode stem (figure 7) and the electrode is tested against a saturated calomel electrode in pH buffers 4.01, 6.86 and 7.41. Electrodes will generally respond immediately with a response closely approximating the theoretical one. If not, the electrodes are examined for air bubbles. When these are found, they are removed by using the microforge wire. The air bubble is made to expand and by careful placement of the microforge wire, the liquid can be distilled down into the tip of the electrode (2). This is continued until the air bubbles float out of the narrow portion of the electrode. Sluggish electrodes are soaked in warm chromic acid (60°C) for a couple of minutes. Electrodes may be left in chromic acid for long periods of time without deleterious effects as long as they are exercised in two buffers before use.

CONCLUSION

Spear-type pH glass electrodes have been constructed with tip diameters down to one micron and sensing lengths of 50 microns. The electrodes are fast-responding with linear and reproducible responses. They can be cleaned in warm chromic acid solution.

BIBLIOGRAPHY

1. Khuri, R.N., Personal Communication 1971.

2. Hinke, J.A.M., Cation-Selective Microelectrodes for Intracellular Use in Glass Electrodes for Hydrogen and Other Cations, George Eisenman, editor, Marcel Dekker, Inc., New York, 1967.

3. Caldwell, P.C., An Investigation of Intracellular pH of Crab Muscle Fibers by Means of Micro-Glass and Micro-Tungsten Electrodes, J. Physiol., 126, 169-180 (1954).

4. Paillard, M., Direct Intracellular pH Measurements in Rat and Crab Muscle , J. Physiol., 223, 297-319 (1972).

5. Caldwell, P.C., Studies on the Internal pH of Large Muscles and Nerve Fibers, J. Physiol., 142, 22-62 (1958).

6. Zorokina, Z.A., Relationship between the Intra- and Extra-Cellular pH, the Rest Potential, and Concentration of Potassium in the Transversostriated Muscular Fiber of the Frog, Cytologia, 3, 49-59 (1961).

7. Lavallee, M., Intracellular pH of Rat Atrial Muscle Fibers Measured by Glass Micropipette Electrodes, Circulation Research, 15, 185-193 (1964).

8. Hirakawa, S., Yoshimura, H., Measurements of the Intracellular pH in a Single Cell of Nitella flexilis by Means of Micro-Glass pH Electrodes, Jap. J. Physiol., 14, 45-55 (1964).

9. Rector, F.C., Carter, N.W., Seldin, D.W., The Mechanism of Bicarbonate Reabsorption in the Proximal and Distal Tubules of the Kidney, J. Clin. Invest., 44, 248-290 (1965).

10. Khuri, R., pH Glass Microelectrodes for in-vivo Applications, Rev. Sci. Instr., 39, 730-732 (1968).

11. Zorokina, Z.A., Measurement of the Activity of Hydrogen Ions Outside and Within the Nerve Cells of Ganglia of Mollusks, Zh. Evol. Biokhim. Fiziol., 1, 341-350 (1965).

12. Carter, N.W., Rector, F.C., Campion, D.S., Measurements of Intracellular pH of Skeletal Muscle with pH Sensitive Glass Microelectrodes, J. Clin. Invest., 46, 920-933 (1967).

13. Carter, N.W., Rector, F.C., Campion, D.S., Seldin, D.W., Measurements of Intracellular pH with Glass Microelectrodes, Fed. Proc., 26, 1322-1326 (1967).

14. Paillard, M., Determination Directe du pH des Cellules Musculaires du Rat, in-vivo et in-vitro, a L'aide de Microelectrodes de Verre, These pour le Doctorat en Medecine, Faculte de Medecine de Paris, France, 1969.

15. Khuri, R.N., Agulian, S.K., Harik, R.I., Internal Capillary Glass Microelectrodes with a Glass Seal for pH, Sodium, and Potassium, Pflugers Archiv., 301, 182-186 (1968).

16. Loewe, R., Linzen, B., V. Stackelberg, W.F., Die Gelosten Stoffe in der Hamolymphe einer Spinne, Cupiennius Salei Keyserling, Z. Vgl. Physiol., 66, 27-34 (1970).

17. Thomas, R.C., Membrane Current and Intracellular Sodium Change in a Snail Neurone During Extrusion of Injected Sodium, J. Physiol., 201, 495-514 (1969).

18. Kessler, M., International Symposium on Microchemical Techniques, Pennsylvania State University, August 1973.

19. de Fonbrune, Pierre, Technique de Micromanipulation, Masson et Cie., Editeurs, Paris 1949.

20. Hebert, N.C., Properties of Microelectrode Glasses in Glass Microelectrodes, edited by M. Lavallee, O. Schanne, & N.C. Hebert, published by John Wiley & Sons, N.Y. 1969.

21. Properties of Selected Commercial Glasses Corning Glass Works, Bulletin B-83.

DISCUSSION

Question: (Zeuthen)

Why do you prefer the Hinke design to the R. Thomas design?
Comment on: time constants
: easiness of fabrication

Answer: (Hebert)

The main difference between the two microelectrodes is that the Hinke design has an exposed sensing tip whereas the Thomas design has a sensing tip which is recessed inside of a glass micropipette. The Hinke-type microelectrode senses on contact with the test medium. The Thomas-type requires that the test medium equilibrate with the distilled water which occupies the space between the sensing portion of the electrode and the open tip of the micropipette.

The response time of the Hinke-type microelectrode is instantaneous, i.e., as fast as the electrometer can respond. The response time of the Thomas-type microelectrode can be several minutes and is depended upon equilibration of two solutions separated by the outer pipette tip having a two micron inside diameter tip.

The only advantage of the Thomas-type design over the Hinke type design is its short sensing length. Theoretically only the fluid which touches the tip-end and diffuses into the sensing chamber will be analysed. The Hinke-type microelectrode has a sensing length of about 50 microns. We have, however, made these to 10 microns on occasion. The sensing length limits the applications of this microelectrode to large cells since the total sensing length of the microelectrode should be in the compartment to be analysed.

We find the Hinke-type microelectrode easier to make. This is likely due to our spending much more time in making this type of microelectrode.

Question: (Whalen)

Have you ever tried saran dissolved in methyl ethyl ketone to cover (insulate) your pH sensitive electrodes? One could do it with a drop and raise the tip of the inverted electrode while gently blowing on the drop. You could be looking in a microscope, coat and recoat and recoat to any thickness always starting at the same point on the electrode leaving it clean to the tip.

Answer: (Hebert)
We have not used "saran" as an insulating material. We have, however, tried many others, such as, silicones and varnishes, and found none to be satisfactory. These materials flake off, absorb solution electrolytes, and do not form good seals with glass. We have never been able to form a reliable insulation with any of these materials.

Question: (Saad)
What is the composition of the weight which you attach to the glass in order to draw the tip to the size you want? Is it glass, also?

Answer: (Hebert)
The small weight is made by glass sealing a platinum wire (25 microns in diameter) loop into the open end of a glass tube (Code 0120). The actual weight varies. We use a weight of about 0.20 grams. Actually for best results, a small weight should be used in making the glass-to-glass seal and a heavier weight used in breaking off the tip. In breaking off the tip of the microelectrode, the heavier the weight, the shorter will be the sensing length.

Question: (Durst)
Is it possible to reduce the length of the pH glass tip by applying a vacuum to the inside of the pH capillary when microforge heating is applied?

Answer: (Hebert)
No, the softening points of the insulating glass (630°C) and the pH glass (655°C) are too close. Applying a vacuum could result in caving in the two glasses. On the other hand, a slight air pressure might help in making the glass-to-glass seal.

Question: (Wright)
How are the electrodes filled with the internal solution?

Answer: (Hebert)
Our pH microelectrodes are filled by boiling in distilled water for about 20 minutes. The electrodes are allowed to cool slightly and the distilled water is displaced with 0.1 N HCl by means of a glass needle attached to a syringe. The electrodes are then ready for testing in about one hour.

Question: (Hollander)
Use of CH_3OH and water aspirator vacuum will fill electrodes, open or closed, with outer tip diameters as small as 350 A°?

Answer: (Hebert)

Filling of electrodes under reduced pressures is certainly recommended for preserving fine tipped micropipettes. However, we have not found it necessary to use this technique with our pH microelectrodes which have tip diameters of 1 to 3 microns.

Question: (Spring)

Is it possible to treat the pH sensitive glass to desensitize it over the shaft, but leave the tip pH sensitive?

Answer: (Hebert)

This technique was used by Carter (U.S. 3,129,160, April 14, 1964) in making pH electrodes. We have made some attempts to desensitize electrode shafts by ion exchange in molten salts without success.

Questions: (Paterson)

1. Can the pH glass microelectrode be inserted into a 23G or lesser size hypodermic needle?
2. Are there difficulties in cleaning such a device?
3. Can the microelectrode be removed from the needle and then reinserted into another needle?

Answers: (Hebert)

The spear-type glass pH microelectrode can be made with a long shank which can be inserted into a hypodermic needle having an inside diameter as small as 100 microns. We have made such electrodes with the sensing part of the electrode resting in the bevel of the needle. These are difficult to clean. Foreign materials have a tendency to accumulate in the bevel.

We prefer to use a sensing bulb-type electrode for this purpose. The pH sensing bulb is blown onto the end of an insulating glass tube using a microforge wire. We make these reproducibly to fit inside of 21 gauge needles. This means that the glass bulb has to be smaller than 0.020 inch. The glass electrode can be removed and placed inside of another needle.

SOME PROBLEMS WITH AN INTRACELLULAR PO_2 ELECTRODE[1]

W. J. Whalen, Ph.D.

Director of Research, St. Vincent Charity Hospital

2351 East 22nd Street, Cleveland, Ohio 44115

The micro O_2 electrode we have developed in our laboratory has been described in detail (4). Briefly, the electrode is made by filling a glass capillary with a molten mixture of Wood's metal and gold. The capillary is then drawn out with heat to a sharp point. The shank has a long taper. After inserting the wire contact the tip is bevelled (45° or less) on a diamond-dust-covered rotor. At the beginning of the bevel the electrode measures 2 - 3μ. The metal in the tip is etched away electrolytically until there is a recess of 20 - 40μ. On the metal in the recess a layer of gold usually 5 - 15μ thick, is plated. The recess is usually filled with collodion to act as a filter for large molecules. The electrode (cathode) is used in the usual way with a Ag-AgCl anode (0.7 - 1.0V).

The construction of the electrodes is extremely difficult. Most persons who have tried to make the electrode have found at some time or other that the glass cracks after the capillaries are drawn out. The incidence of this difficulty is less if the filled capillaries are allowed to remain on the hot plate for a few hours prior to pulling. A second major problem is in the etching process. The recess must be absolutely clean prior to plating. We can offer no help here except to say it is an art which takes time to learn.

We have recommended calibration of the electrodes *in situ* if at all possible (3). Not only is there less chance of breaking

[1] This work has been supported in part by grant nos. HL 11906 and HL 12703 from the U.S.P.H.S.

the electrode (in the process of transferring it to a calibration chamber) but since the polarizing voltage is not removed, the calibration is more certain. Solution equilibrated with air at the temperature of the tissue to be explored is simply run through or over the preparation when the electrode is to be withdrawn. For the zero calibration we use a slice of rat brain adjacent to the tissue under study. The zero level however does not change much, if any, if the air value remains constant.

Another problem associated with the use of the electrode is the question of the precise location of the tip when it is deep in the tissue. One can use the cell membrane potential as an indicator (2), but the response time of the amplifier must be very rapid, which makes it extremely sensitive to movement, line surges, etc. We have found that in muscle the tip is in a cell about 70% of the time in random penetrations, and so usually make no attempt to record potentials. (1)

I have already alluded to another problem, that of movement artifacts. These stem largely from the fact that the currents measured are so small (air values 1 to 8 X 10^{-12} Amps.)

The most puzzling problem, if indeed it is a problem, is the fact that some of the electrodes show a reversed current flow when the tip is in a low O_2 environment. The extreme example came from an electrode we recently tested which was negative in physiological saline equilibrated with 5% O_2, even at an applied voltage of 0.9V (in the opposite direction, of course). At lower voltages the negativity is accentuated. In spite of the negativity these electrodes calibrate linearly with the Astrup-analyzed PO_2, at least up to the air value, which is as far as we have gone. In the past we have not used these negative electrodes in our reported work, but perhaps we have been over-cautious. I mention it here in the hope that someone will have an explanation for this phenomenon.

1. Whalen, W. J., D. Buerk, and C. Thuning. Blood flow-limited oxygen consumption in resting cat skeletal muscle. Am. J. Physiol. 224: 763-767, 1973.

2. Whalen, W. J. and P. Nair. Intracellular PO_2 and its regulation in resting skeletal muscle of the guinea pig. Circ. Res. 21: 251-261, 1967.

3. Whalen, W. J. and P. Nair, R. A. Ganfield. Measurements of oxygen tension in tissues with a micro oxygen electrode. Microvasc. Res. 5: 254-262, 1973.

4. Whalen, W. J., J. Riley, and P. Nair. A microelectrode for measuring intracellular PO_2. J. Applied Physiol. 23: 798-801, 1967.

DISCUSSION

Questions: (Durst)

1. How is the oxygen microelectrode fabricated?
2. Would mercury plated onto the tip rather than gold increase the hydrogen overvoltage and thereby increase the range of the plateau?

Answers: (Whalen)

1. Recessing is done by applying 10-20 V (electrode neg.) while the tip is in "cheap" gold plating solution. Either the metal is removed, electrolytically or simply by the heat generated at the tip. I don't know which.
2. Sounds like a good idea!

Question: (Zeuthen)

Don't the steep gradients in pO_2 across the vivo brain correspond to capillary distribution?

Answer: (Whalen)

Yes, we can see these gradients when we pass the electrode over the surface of the brain, when N_2-equilibrated solution is flowing over it.

Question: (Brown)

What is the effect of intracellular voltage on the 1-V relationship on output of the oxygen electrode?

Answer: (Whalen)

It does have an effect, i.e., a drop in "current" transiently as the tip enters the cell, and only if there is a flat plateau in the current-voltage curve do we consider it an absolute measurer of pO_2. It is especially troublesome in the beating heart where the action potential (intracellular) varies with the applied voltage.

Question: (Schultz)

I found your remarks on voltage reversal of the electrode at zero oxygen concentration most interesting. We have also experienced this phenomenon in vivo in a rabbit with our O_2 (Ag) - Pb membrane cells embedded subcutaneously in a "sterile" abscess where the pO_2 approaches zero.

Answer: (Whalen)

That is interesting, and comforting that others have seen a similar phenomenon. It does seem to occur most often when the electrode tip is in tissue rather than in N_2 equilibrated solution.

SOME PROBLEMS WITH THE ANTIMONY MICROELECTRODE

R. Green and G. Giebisch
Department of Physiology
Yale University School of Medicine

New Haven, Connecticut 06510

INTRODUCTION

Some twenty or thirty years ago electrochemists investigating the antimony electrode with a view to using it in pH determinations found it to be unstable, and the results not reproducible. The changes in pH evidently altered the characteristics of the physical state of the antimony. The electrode potential was linear with increasing pH up to about pH 7.0. Above this value there was some reduction in the slope. Different investigators did not agree as to the pH at which this change in slope occurred, but all agreed that it did occur (see 2,8). The major disadvantage was that antimony electrodes had to be calibrated frequently, and as a consequence they were not popular.

As renal physiologists developed the technique of micropuncture, investigators sought an electrode to measure pH which did not allow carbon dioxide to equilibrate with the bicarbonate in vitro as happens with the quinhydrone electrode. An alternative was the ultramicroglass electrode (6), but the technical difficulties in its fabrication (1) stopped many potential users. In 1968, Malnic and Vieira (9) introduced the micro antimony electrode into micropuncture research and have used it extensively since then (5). Because of increasing usage of antimony microelectrodes, a survey of some hitherto undescribed difficulties we encountered in its use should be most helpful to other investigators.

Results and Discussion

The objective of this procedure is to measure titratable acidity. This requires using two antimony electrodes and two reference electrodes as described by Karlmark (3, 4). To avoid the problem of having four micromanipulators around a sample plate, double barrelled pipettes were fabricated. One barrel was filled with antimony and the other with a reference solution, in this case 2M potassium chloride and 0.5M potassium nitrate in 2% agar. This reference solution minimizes potential changes caused by the varying anionic composition and concentration in the test solutions (1). These electrodes were used throughout with no cross coupling between the barrels of single pipette.

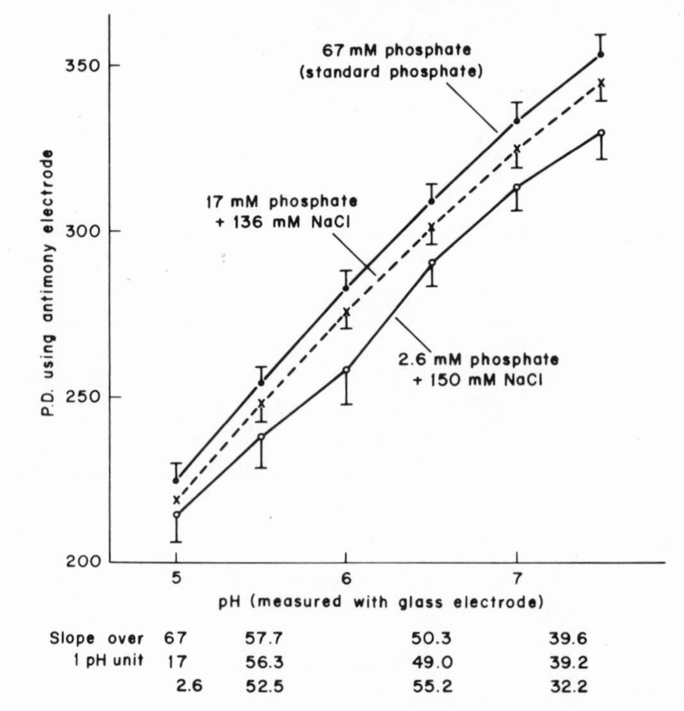

Fig. 1. Potential developed in antimony electrodes compared with pH measured with glass electrode for three sets of standard solutions containing 67, 17 and 2.6 mM phosphate buffers. Bars are $\pm$ 1 S.E. Slopes for the standard curves are shown below the graph.

Effects of Varying Phosphate Concentration

Calibration of the electrodes immediately demonstrated one

difficulty. With standard phosphate buffers made up after Sørensen (7) (but using only the sodium salts) and having a final phosphate concentration of 67 mM, the calibration curve was similar to those previously described. (See Figure 1). The pH of the standards were measured with a glass electrode (Radiometer Copenhagen Model 27) which was routinely checked against two other glass electrodes, and the potential of the antimony electrode was measured on a voltmeter specifically designed in this department by Mr. Harry Fein. Over the lower pH values the response of the electrode was linear with a slope of 57 mV per pH unit, almost approaching the theoretical value. Above pH 6.5 the slope decreased gradually, a finding in agreement with those of Malnic (5).

Almost by accident, it was discovered that the pH of a solution containing 17 mM phosphate was not the same when measured with the glass electrode and with antimony electrodes calibrated with 67 mM phosphate standards. This difference persisted after rechecking the electrodes and was a constant feature of all the electrodes tested. The difference was still greater in 2.6 mM phosphate buffer solutions. This latter phosphate concentration is more analogous to the fluid in renal tubules. Consequently it was possible to construct a calibration curve for each of the three total phosphate concentrations. Figure 1 shows these three curves and the slope per pH unit. Since, the pH was measured with a macroglass electrode, dilution effects and alterations of ionic strength can not be invoked to account for the difference. Throughout this investigation it has been assumed that the pH, as measured with the glass electrode, is the true expression of hydrogen ion activity.

The consequence of these three calibration curves is shown in Figure 2 where the pH of phosphate solutions measured with a glass electrode is compared with the apparent pH of the same solutions measured with antimony electrodes previously calibrated with 67mM phosphate buffer standards. If, for example, one has a solution containing 2.6 mM phosphate with a true pH of 7.40, the reading with calibrated antimony electrode, will be an apparent pH of 6.80. It would therefore appear that to get true pH value with the antimony electrode, the buffers used in the calibration process should have the same phosphate concentration as the unknown solution.

Effects of Varying Ionic Strength

The data also showed that alterations in the ionic strength of the solution by addition of saline, alter the potential relatively more when it is measured with an antimony electrode, than when it is measured with a glass electrode. For example, increasing the ionic strength of 17mM phosphate buffer solution by adding 0-200mM of sodium chloride resulted in a decrease of 0.38 pH units with a glass electrode, but 0.54 units with the antimony electrode.

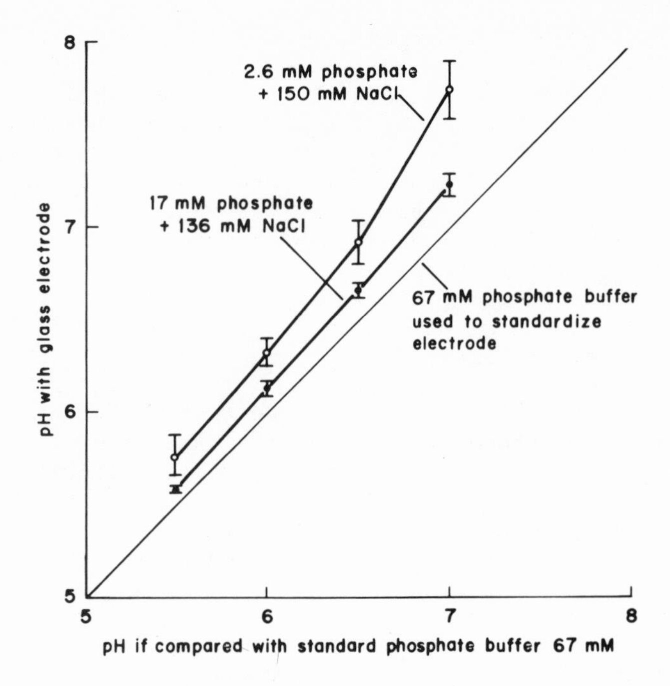

Fig. 2. True pH of phosphate containing solutions measured with glass electrode plotted against the apprent pH when measured with antimony electrodes calibrated with 67mM phosphate standards.

It would appear that antimony electrodes can be used reliably in experiments on proximal renal tubules, where there is little change in the phosphate buffer concentration and probably no change in ionic strength, providing that the phosphate buffers and ionic strength of the standards match the unknown fluid. Questionable data should be expected with antimony microelectrodes in distal tubules, collecting ducts and in final urine, where ionic composition and buffer capacity vary widely.

Effects of Varying Bicarbonate and Acetate Concentrations

Although it is claimed that the reference electrode minimizes potential changes due to changes in anionic composition (1), it was essential to verify this claim since we intended to expose the electrode to solutions which varied in their acetate and bicarbonate concentrations from 0-25mM. Furthermore, Karlmark (4) had previously reported an apparent 0.16 unit reduction of pH whenever bicarbonate was present.

Bicarbonate or acetate was added to solutions containing 17mM phosphate buffer. In each case the concentration of bicarbonate and acetate was 25mM, equilibrated at a different pH. The ionic strength was kept constant by addition of appropriate amounts of sodium chloride. The pH of the resulting solutions were measured with a glass electrode and with antimony electrodes calibrated with both 17mM phosphate and 67mM phosphate standards. The results (A.M. ± S.E.) are presented in Table I. Three different bicarbonate solutions were measured with seven different antimony electrodes. The four acetate solutions were measured with eight antimony electrodes. It is obvious from the data that bicarbonate does not cause a discrepancy in pH reading if the antimony electrode is correctly calibrated in standards with the same total phosphate concentration. This finding is also applied equally for acetate. Other anions should be checked before antimony electrodes are used in solutions which contained them.

EFFECT OF ANIONS

		17mM Phosphate Standard	67mM Phosphate Standard
	η	Δ pH	Δ pH
Bicarbonate	21	+0.003	-0.255
	(3x7)	±0.022	±0.044
Acetate	32	+0.025	-0.113
	(4x8)	±0.009	±0.021

TABLE I. Mean deviation of pH obtained with appropriate phosphate standarized antimony electrodes from the pH measured with a glass electrode, for the same solutions containing bicarbonate and acetate. (Figures in parentheses indicate 3 solutions of bicarbonate measure with 7 electrodes, and 4 solutions of acetate measured with 8 electrodes).

Effects of Varying The Volume of Test Fluid

All the foregoing experiments were done in beakers containing upward of 3 ml of fluid. When titrations are performed on actual samples, only 5-10 nl of fluid are available. Testing with varying size drops showed that the voltage reading from the antimony electrode placed in drops of different size varied, even when the drops were from the same parent solution. The potential of drops of different sizes from the same solution were measured with five different antimony electrodes, and the mean recorded potential was plotted against time (Figure 3). A systematic examination of the data revealed the following interesting relationships: Basically, as the size of the droplet decreased, the initial potential reading and the final reading both decreased. In the solution all potentials decreased with time over the first minute, and most continued to decrease after that time. At variance with these generalizations was the behavior of very small droplets (represented by the 1.8 nl drop) which decreased initially but then increased. The reason for this curious behavior is a mystery. It was not due to droplet shrinkage or movement of the droplet up the tip of the measuring electrode.

The data showed that a linear relationship does not exist between the pH determined with the antimony electrode and that with

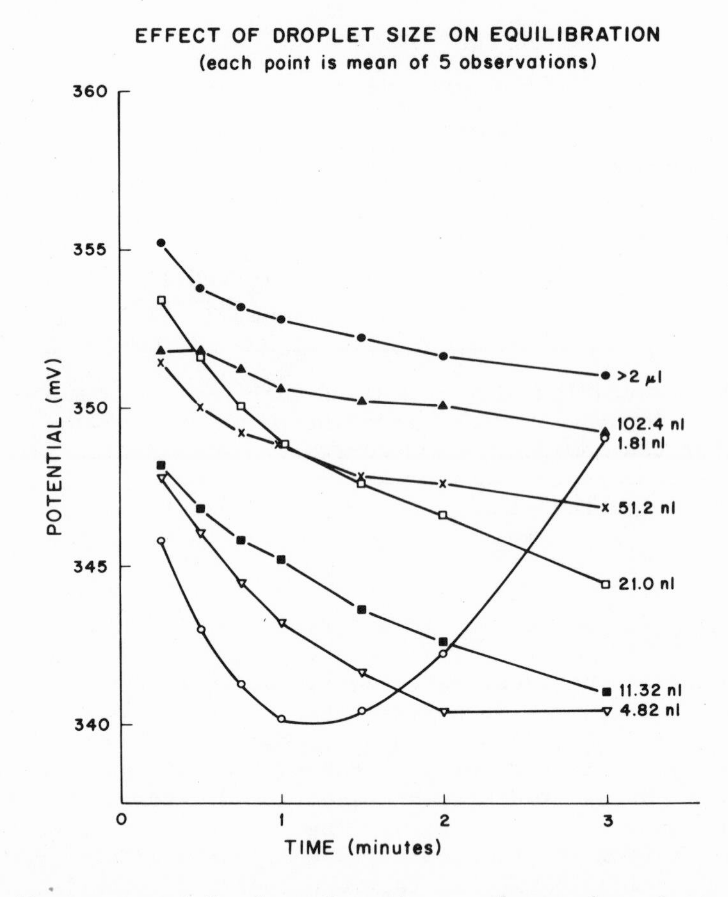

Fig. 3. Mean potentials from 5 antimony electrodes plotted against time for different size drops of the same solution.

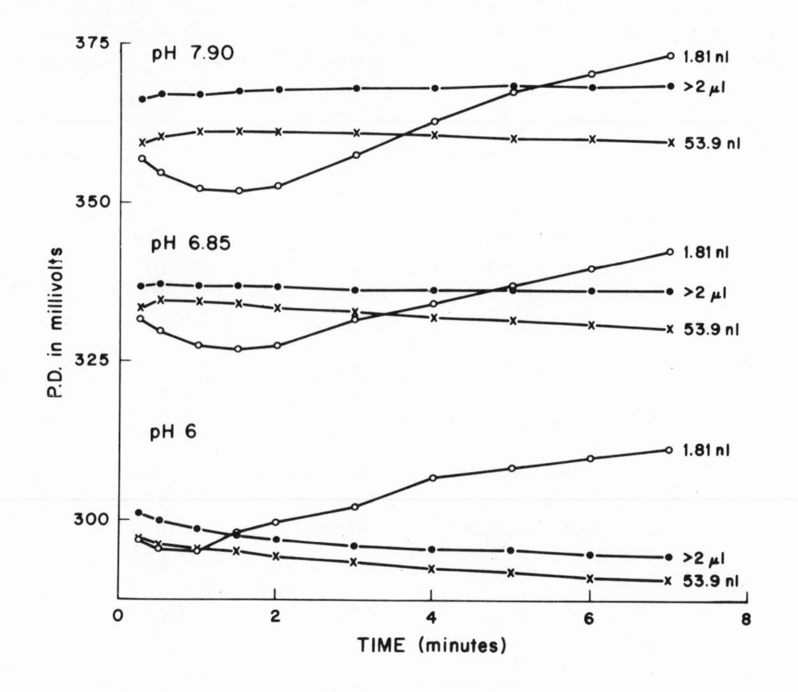

Fig. 4. Potential plotted against time for 3 different size drops at 3 different pH vales.

the glass electrode at higher pH values. The previous experiment was repeated at differing pH values as shown in Figure 4. The potential of three different sizes of droplet at 3 different pH values were measured with the antimony electrode and the readings plotted against time. After about 1-2 minutes the potential of the larger drops became fairly stable. The differences in absolute potential between the 50 nl drop and the larger drops were more marked at higher pH values where the non linearity of the calibration occurs. However, the anomalous behavior of the very small drops occurred at all pH values.

Hence it would appear logical to measure standards and unknowns, in vitro, in drops of equal size which are preferably not too small.

A related but more specific problem occurs when titratable acidity is measured. As described by Karlmark (3), when a measuring antimony electrode and its reference electrode are placed in a drop of unknown solution and a second antimony electrode and its reference electrode placed in the same drops are used to pass current, the current liberates hydroxyl ions from the antimony and make the solution more alkaline. The potentials of both the standard and the test solution drift with time, as is represented diagrammatically in Figure 5. The end point of the titration is the reading of a standard drop with a pH of 7.4. It too has its potential measured with time and the time curve preferably is checked in a second drop of standard. In this manner the potentials at certain specified times are determined. The test drop then has its potential measured for 30 seconds or so after which titration is begun. The titration is to the previously determined drifting time end-point. Great care has to be taken to prevent overtitration. After titration has ceased, the drift of the test droplet should be the same as the drift of the standard.

Summary

In summary, when using antimony microelectrodes the standards should have the same phosphate concentration and total ionic strength as the unknown and all measurements should be made in drops of equal size. All readings are time dependent, and in small drops, the reading should be taken after a suitable time interval, usually 1-2 minutes.

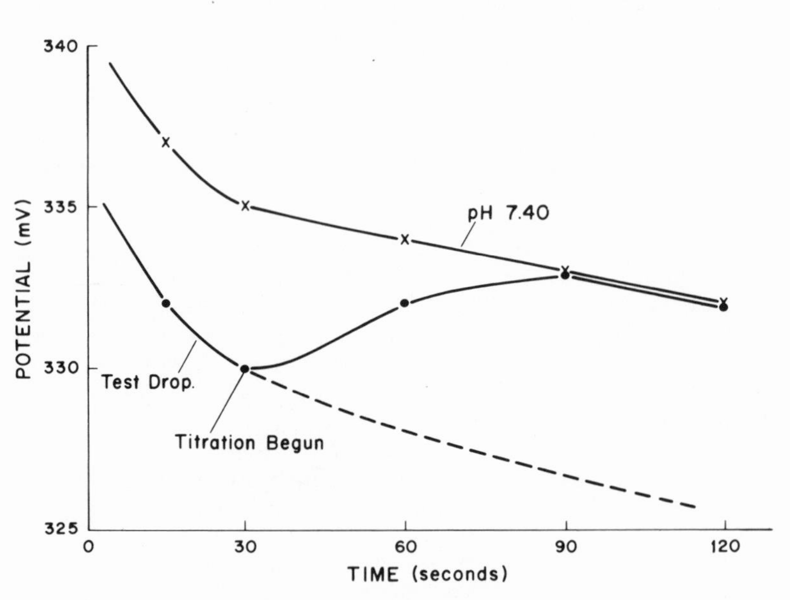

Fig. 5. Diagrammatic representation of measurement of titratable acidity. Test drop is titrated back to standard, a drop at pH 7.4.

TITRATABLE ACID (mEq/L)					
Actual	0.25	0.50	1.0	2.50	5.0
Measured	0.28 ±0.02	0.50 ±0.02	1.04 ±0.02	2.61 ±0.10	4.72 ±0.21
n.	8	8	8	6	6

TABLE 2. Titratable acidity measured as described in text compared with the actual concentration of acid in a standard buffer solution. (± S. E.)

In spite of the variations evidently inherent in the antimony microelectrode, it is still possible to use the antimony microelectrodes reproducibly as is shown in Table 2. In the table the actual concentrations of acid are compared with the analyzed "titratable acidity" of 5 nl samples, and the agreement is quite close. The coefficient of variation is between 5-10% which is as good as most methods used in micropuncture work.

ACKNOWLEDGEMENTS

Thanks are due to Mr. Harry Fein who developed the constant current passing device used here for measuring titratable acid.

REFERENCES

1. Carter, N.W., The Production and Testing of Double Barrelled pH Glass Microelectrodes for Measurement of Intratubular pH, Yale J. Biol. and Med. 45:349-355 (1972).

2. Elwakkad, S.E.S., The Electrochemical Behaviour of the Antimony Electrode, J. Chem. Soc. London, 2894-2896 (1950).

3. Karlmark, B. An Ultramicro Method for the Separate Titration of Hydrogen and Ammonium Ions, Pflugers Arch. 323:361-365 (1971).

4. Karlmark, B. Net Hydrogen Ion Secretion by the Proximal Tubule of the Rat, M.D. Thesis, University of Uppsala (1972).

5. Malnic, G. and Vieira, F.L. The Antimony Microelectrode in Kidney Micropuncture, Yale J. Biol. and Med. 45:356-367 (1972).

6. Rector, F.C. Jr., Carter, N.W., and Seldin, D.W., The Mechanism of Bicarbonate Reabsorption in Proximal and Distal Tubules of the Kidney, J. Clin. Invest. 44:278-290 (1965).

7. Sorensen, S.P.L. Uber die Messung und Bedeutung der Wasserstoffionen Konzentration bei biologischen Prozessen, Ergebn Physiol. 12:393-532 (1912).

8. Tourky, A.R. and Mousa, A.A., Studies on Some Metal Electrodes III. Does the Antimony Electrode Behave Simply as a Metal-Metal Oxide Electrode in Air? J. Chem. Soc. London. 752-755 (1948).

9. Vieira, F.L. and Malnic, G. Hydrogen Ion Secretion by Rat Renal Cortical Tubules as Studied by an Antimony Microelectrode, Amer. J. Physiol. 214:710-718 (1968).

DISCUSSION

Question: (Puschett)

Could you describe your in vitro system a bit more?

Comment: (Puschett) We have now had a fair amount of experience with an ultramicro technique for measuring pH both with antimony and glass microelectrodes. We noted that there was very little change in the pH measured with the antimony electrode as time elapsed, once equilibration had occurred, when we used electrodes that were 3-4 weeks old as compared to using those made the same day. We presumed this was due to the fact that the electrode had, in fact, become somewhat oxidized in the interim.

Answer: (Green)

(to question) Recently made (within 1-2 weeks) double-barrelled electrodes were inserted into 5-10 nl droplets under oil equilibrated at 37°C with 20% O_2 80% N_2 mixture and saturated with water.

(to comment) This may be true. Although with drops of the order of 25-35 nl as you have used, I would expect that drift would have stopped after 1-2 minutes anyway. Our problem is with the smaller droplets of the order of 5-10 nl.

Question: (Malnic)

What do you think may be the reason for pH changes in small droplets under oil? We have had the problem of acid contamination of oil, which could be removed by washing the oil with distilled water, after which stable measurements were obtained.

Answer: (Green)

It is possible that the oil plays some part but our oil does not appear to be acid. It seems more likely that the change is due to some dissolution of the antimony trioxide in the droplets.

Question: (Durst)

What is the possibility that the anomalous drift observed in the nanoliter samples is caused by dissolution of the antimony oxide producing a change in pH in the basic direction?

Answer: (Green)

The possibility that this occurs is quite good. However, from our experiments we could not determine whether this possibility is the correct one or not.

Question: (McDonald)

Is the potential drift recording of pH with antimony electrodes caused by a change in oxygen concentration of the droplet? Investigators report that pH antimony electrodes are sensitive to the PO_2.

Answer: (Green)

Some investigators report effects due to altering PO_2 and some do not. However, to avoid any effects of this nature all our experiments were carried out at 20% O_2 concentration. Hence, the drift cannot be due to changing O_2 concentration.

II. Intracellular Applications

IONIC ACTIVITIES IN IDENTIFIABLE APLYSIA NEURONS

Arthur M. Brown, M.D., Ph.D. and Diana L. Kunze, Ph.D.
Dept. of Physiology
University of Texas Medical Branch
Galveston, Texas 77550

This paper reports the values of K^+, Na^+ and Cl^- activities measured with ion-selective microelectrodes in certain identifiable neurons in the abdominal ganglion of Aplysia californica (nomenclature according to Frazier, et al., 1967). These measurements allow calculation of the equilibria potentials of the three ions thought to be most important in regulating intracellular voltage. The ion-selective microelectrode technique makes such measurements possible in small cells; moreover these values can be compared amongst neurons which are anatomically and functionally distinct.

METHODS

In Vitro Methods

Liquid ion-exchanger type microelectrodes for measuring Cl^- and K^+ activities were fabricated according to the method adumbrated by Orme (1968) and improved by Walker (1971). The liquid ion-exchangers used were obtained from Corning Glass Works, Scientific Instrument Department. The Na^+ glass microelectrode was fabricated according to the method of Thomas (1970) as modified by Brown and Brown (1973). The electrochemical response curves and ion-selectivities of these microelectrodes have been reported (Orme, 1968; Brown, Walker and Sutton, 1970; Brown and Brown, 1973). These papers should be referred to.

In Vivo Methods

A pair of glass micropipettes (tip diameter 1μ, resistances 3-4 MΩ) filled with a variety of solutions, generally 3.0 M KCl, were used to record the membrane potential, (E_M) by inserting one pipette in the neuron and using the other as a reference electrode. These were connected to an electrometer with capacity neutralization. Another micropipette (< 1 MΩ resistance) was inserted for current injection in voltage- or current-clamp experiments. The clamping circuit was described by Eaton (1972). One or more ion selective microelectrodes were inserted into the same neuron. The ion selective electrodes were connected to a varactor bridge operational amplifier having an input impedance of 10^{14} ohms. The electrometer output was displayed on a digital voltmeter or recorded differentially using E_M as the other input. Voltages and currents were displayed on a Tektronix 565 oscilloscope and recorded on a penwriter, (Russell & Brown, 1972 a & b).

Ionic activities were measured in R_2, R_{15} and the left upper quadrant cells L_{1-4} and L_6. Measurements were often made upon several cells from the same animal. Cells L_1–L_4 and L_6 have been grouped together for the following reasons: 1) they are pacemaker cells, 2) they receive a common inhibitory input from interneuron L_{10}, and 3) they respond similarly to iontophoretic application of acetylcholine (ACh.).

To ensure that all electrodes were in the same cell constant current pulses were passed across the cell membrane. Equal voltage deflections on all intracellular microelectrodes ensured that such was the case.

The measurements for K^+ and Cl^- are accurate to 5% over the range of values found (Russell & Brown, 1972 a & b); they are accurate to about 7% for Na^+ (Russell & Brown, 1972 c). No correction for Na^+ was made in the K^+ determinations since the omission leads to less than a 1% error, (see Table 1). The correction for K^+ was essential for the Na^+ measurements despite the 50:1 selectivity of the Na^+ electrode (Table 1) because the error due to omission would amount to about 20%.

The reversal potentials E_{ACh} for the effects of ACh on R_2 and the left upper quadrant follower cells, and for synaptic potentials evoked by L_{10} on the follower cells were determined for comparison with the appropriate ionic equilibria potentials.

RESULTS

The time constants for the K^+, Na^+ and Cl^- microelectrodes used in these experiments were about 1 sec, 2 min and 2 sec, respectively. However, the time required for reaching steady values intracellularly are of the order of 5 min and readings were therefore generally taken over a 30 min period. The longer time required to reach intracellular steady state values may have been due to changes in intracellular activities as a result of leakage following electrode insertion, or it may represent intracellular damage around

TABLE I

a^i_K's and a^i_{Cl}'s for three groups of *Aplysia* neurons

	a^i_K (mM)	n	a^i_{Cl} (mM)	n	a^i_{Na} (mM)	n	E_K (mV)	E_{Cl} (mV)	E_{Na} (mV)	E_M (mV)	n
R_2	165 ± 3.75	36	37 ± 0.8	120	18.8 ± 1.7	9	$-79.5 \pm .55$	$-56.2 \pm .50$	$+73.7 \pm 2.0$	-49.5 ± 0.4	130
L_1-L_6	142.6 ± 3.2	22	$34.5 \pm .74$	27	21.2 ± 1.0	22	$-75.7 \pm .56$	$-56.9 \pm .51$	$+70.6 \pm 1.1$	-43 ± 1.0	60
R_{15}	$137.9 \pm .85$	16	42.2 ± 1.36	17	23 ± 2.0	6	$-75.2 \pm .85$	$-52.4 \pm .61$	$+68.6 \pm 2.2$	-45.5 ± 2.8	35

a^o_K = 7 mM

a^o_{Na} = 350 mM

a^o_{Cl} = 349 mM

n is the number of cells studied in each group

Values are mean $\pm$ S.E.M.

the tips of the microelectrodes, or some combination of the two. (see discussion).

Ionic Activities in Specific Cells

a_K^i and a_{Cl}^i for the three cell types (Figure 1) are recorded in Table 1. E_K for individual cells ranged from 25-30 mV

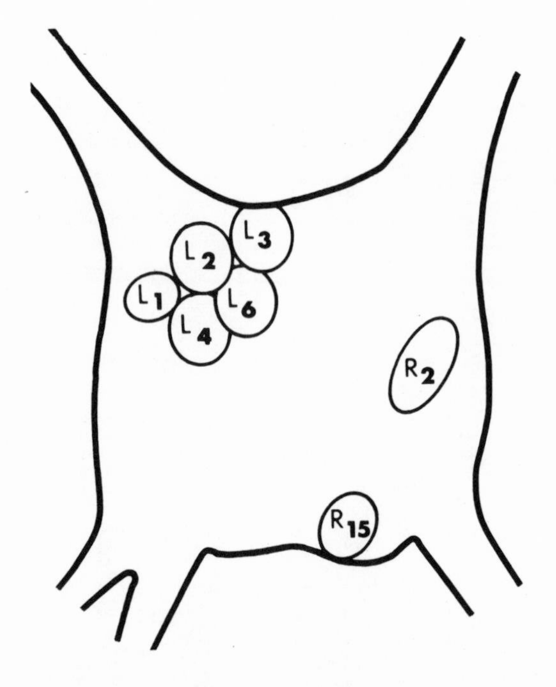

Figure 1. Dorsal surface of the abdominal ganglion of Aplysia californica and the location of specific cells (L_1-L_6, R_2, R_{15}) used in the study. L_{10} (not shown) is located on the ventral surface of the ganglion.

more negative than E_M (Table I). E_{Cl} was also more negative than E_M on the average (Table I) but in six giant cells, E_{Cl} was less negative than E_M. Because of the variation amongst different abdominal ganglia, measurements of a^i_K for each of the three cell types were made in each of five ganglia. A similar study of a^i_{Cl} was done. a^i_K was significantly higher in R_2 than in the other groups ($p < .03$) and a^i_{Cl} was significantly lower ($p < .05$) in the $L_1 - L_6$ group. These results also apply to the large series shown in Table I. The values of a^i_{Na} were not significantly different amongst the three cell types; E_{Na} is much more positive than E_M, which is not surprising. The values for K^+, Cl^- and Na^+ activities are similar to those reported in other series (Russell & Brown, 1972 a, b, & c, Brown & Brown, 1973).

The Effects of ACh and Their Relation to a^i_{Cl} and a^i_K

1. E_{ACh} and E_{Cl} in the giant cell. Figure 2A shows the changes in the membrance potentials of R_2 when ACh was added to the perfusate. The change in the membrane potential, i.e., present potential minus the potential obtained after the addition of ACh, are plotted against preset potential in Figure 2B. The present potential which does not change on the addition of ACh is a measure of E_{ACh}. E_{ACh} measured in this cell (Figure 2B) was -51mv. In five cells the mean value for E_{ACh} was -51.9±0.59 mv (Table 1) calculated from the directly-measured a^i_{Cl} of these cells.

R_2 usually responds to ACh with a hyperpolarizing potential, the so-called H response (Gerschenfeld and Tauc, 1961). However, E_M may be more negative than E_{Cl} and a depolarizing or D response would be anticipated in this instance. Figure 3 shows an experiment in which E_M changed spontaneously over a period of several hours from -41 mV in the first trace (taken one hour after impalement) to -61 mV in the second trace (taken 3 hours later). ACh elicited an H response when E_M was -41 mV and a D response when E_M was -61 mV. We cannot account for the different time courses of the two responses. We have found that a^i_{Cl} did not vary over several hours when it was continuously recorded in these cells, although spontaneous changes in E_M occurred. (Russell & Brown, 1972 a & b).

2) E_{ACh}, E_{Cl} and E_K in L cells. The L cells always respond to ACh with a hyperpolarizing response. The response to iontophoretic application of ACh was easily separable into an early and a late component by suitable polarization of the membrane (Figure 4) as has already been reported (Kehoe and Ascher, 1970; Pinsker & Kandel, 1969; Kunze & Brown, 1971). The reversal potentials for the early (E_{ACh}) and late (E'_{ACh}) components were -57 mV and -76 mV (Table 2). These values were almost identical with the E_{Cl} of -56.9 mV and the

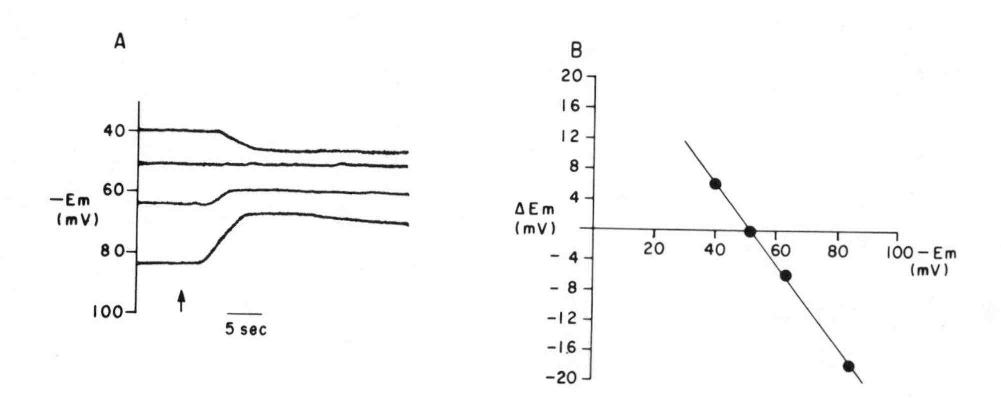

Figure 2. (A) Change in E_M of the giant cell R_2 when ACh (10^{-5}gm/ml) is added at the arrow to the perfusate. The cell was at its resting potential of -40 mV in the top trace. (B) The change in E_M plotted against the preset E_M gives a reversal potential of -51 mV.

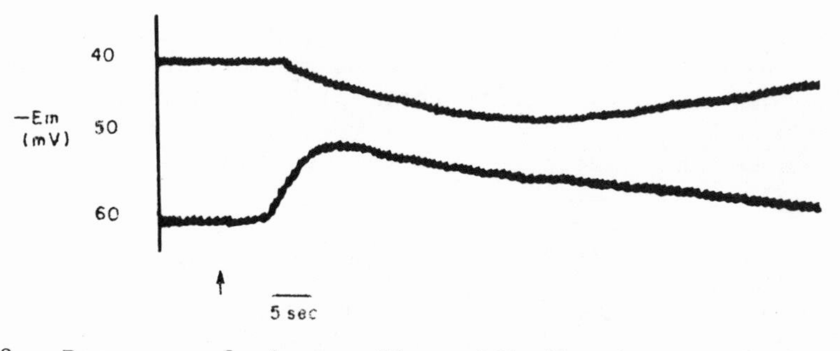

Figure 3. Response of giant cell to ACh (at the arrow) when E_M was -41 mV (upper trace) and three hours later when E_M was -61 mV.

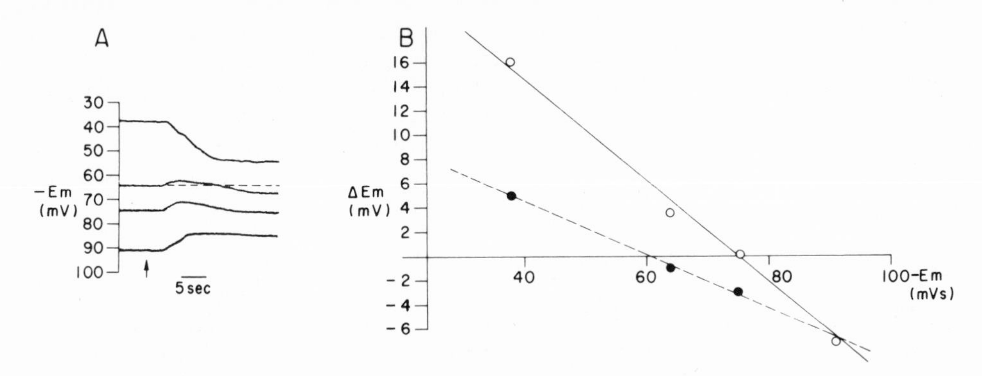

Figure 4. Effect of ACh (.3 M) iontophoretically applied to the soma of L_3 at the arrow (A). The upper three traces show an early component, which reverses at -60 mV (closed circles in B). The late component, taken as the change in E_M 20 sec after application of ACh (open circle B) has a reversal potential of -76 mV.

TABLE 2

Reversal potentials for ACh and synaptic action on *Aplysia* neurons

	E_{ACh} (mV)	n	E'_{ACh} (mV)	n	E_{IPSP} (mV)	n	E'_{IPSP} (mV)	n
R_2	-51.9±.6	5						
L_{1-4} & L_6	-57±1	3	-76±1	9	-56±1	6	-81±.9	6

Values are mean ± S.E.M.

E_K of -75.7 mV measured in these cells (Table 1). Thus

$$E_{ACh} = E_{Cl} \quad (1)$$

and

$$E'_{ACh} = E_K \quad (2)$$

for these cells as shown by Kunze & Brown (1971).

Synaptic action and E_{Cl} and E_K in L cells.

As shown in Table 1 and 2 and Figure 5,

$$E_{IPSP} = E_{Cl} \quad (3)$$

The "late IPSP" was elicited following repetitive stimulation of L_{10} and was more easily measured after the "early IPSP" was blocked by d-tubocurarine (10^{-4}gm/ml) (Figure 6). E'_{IPSP} was -81 mV (Table 2) which was significantly more negative than E_K (Table 1); possible explanations for these differences will be discussed.

DISCUSSION

a^i_K's, a^i_{Cl}'s and a^i_{Na}'s in identifiable Aplysia neurons.

In all cells studied, a^i_K was very much greater than would have been anticipated from a passive distribution (Table 3). Thus, K^+ is actively transported into these neurons. Similar results were

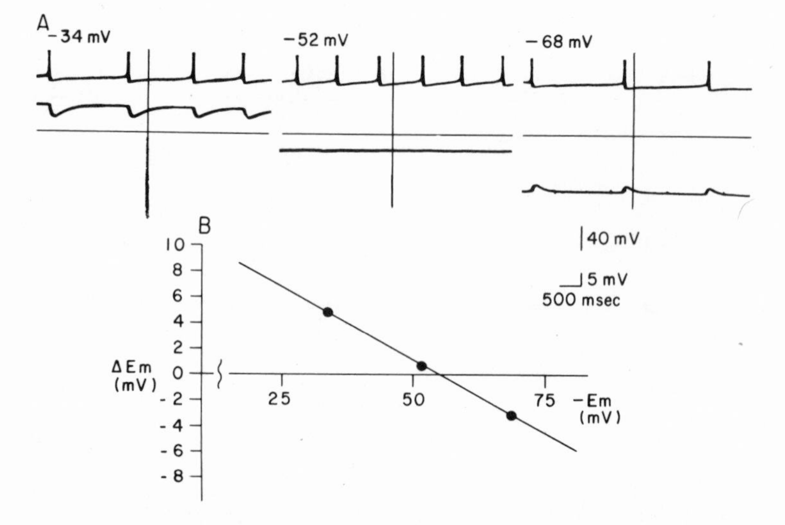

Figure 5. (A) Spontaneous discharge in L_{10} (upper trace) produces an inhibitory synaptic response in L_5 (lower trace) which can be reversed as the E_M of L_5 is hyperpolarized to a level more negative than -54 mV. (B) The amplitude of synaptic response is plotted against the preset E_M of the postsynaptic cell to obtain a reversal potential of -54 mV.

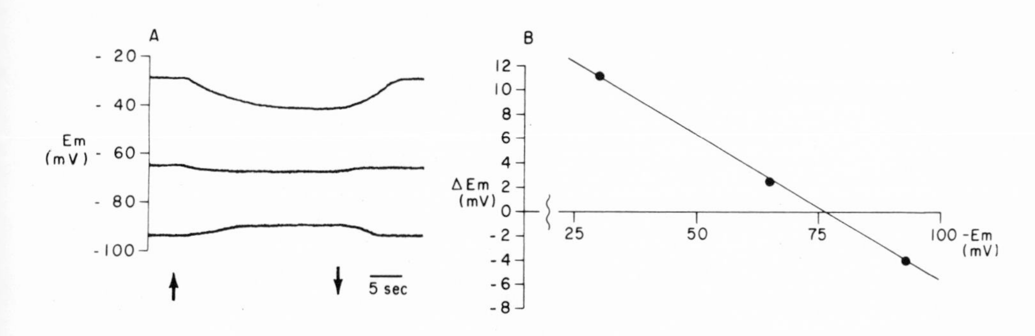

Figure 6. In a curarized preparation which blocks the early IPSP, stimulation of L_{10} at 15/sec between arrows produces a late synaptic potential in L_3 which in this cell reverses at -76 mV.

reported by Sorokina (1966) who used an ion sensitive glass microelectrode to measure a^i_K in neurons of Helix pomatia and Planorbis corneux. Taken collectively, a^i_{Cl} was too low for passive distribution if the cells were in a steady state, which may indicate that Cl^- is actively transported out of these neurons. However, in six giant cells, a^i_{Cl} was greater than that expected (Brown, Walker & Sutton, 1970), which may indicate active transport into these cells. Similar variations in a^i_{Cl} in molluscan neurons were reported by Kerkut and Meech (1966) and Walker and Brown (1970) and were suspected from indirect evidence by Chiarandini, Stefani and Gerschenfeld (1967). Among the three groups of cells, the giant cell, R_2, has a significantly greater a^i_K than the "bursting" cell, R_{15}, or the L cells, while the L cells have a significantly lower a^i_{Cl} than R_2 or R_{15}. The reasons for these differences are unknown.

Na^+ activities were equally low amongst the three neuronal groups.

The intracellular concentrations of Na^+ and K^+ determined by an atomic absorption spectrophotometer were averaged for 400 neurons of the Aplysia abdominal ganglion (Sato, et al., 1968). The values were 232mM for K^+ and 67mM for Na^+. Taking our data on activities, we calculate activity coefficients γ_K of .6 to .7 and γ_{Na} of .24 to .34. The activity coefficients for K^+ as well as Na^+ and Cl^- in seawater are similar to intracellular γ_K suggesting that K^+ is free in the cytoplasm. However intracellular γ_{Na} is very much lower and most of the intracellular Na^+ is therefore probably bound.

TABLE 3

Comparison in three groups of *Aplysia* neurons of a_K^i and a_{Cl}^i measured directly and calculated assuming passive distribution of these ions

	Directly-measured a_K^i (mM)	Calculated a_K^i (mM)*	Directly-measured a_{Cl}^i (mM)	Calculated a_{Cl}^i (mM)*	Directly-measured a_{Na}^i (mM)	Calculated a_{Na}^i (mM)*
R_2	165	50	41	48	19	2547
L_1-L_6	142	38	35	63	21	1929
R_{15}	137	43.2	42	56	23	2173

*Calculated from the Nernst equation assuming that $E_K = E_{Cl} = E_M = E_{Na}$

ACh effects, synaptic action, and their relation to a^i_{Cl} and a^i_K.

For the giant cell, R_2

$$E_{ACh} \simeq E_{Cl}$$

The simplest explanation for this finding is that the receptors for the ACh effect are located electrically at or near the cell body where a^i_{Cl} was measured. While its functional significance is unknown, it is clear that ACh produces a great increase in Cl permeability (Gerschenfeld and Tauc, 1961), E_M moving towards E_{Cl}. E_{Cl} was constant over several hours of continuous recording, but E_M varied. Thus, a D or H response to ACh was obtained, depending upon whether E_M was less negative or more negative than E_{Cl}. This confirms directly the idea proposed initially by Chiarandini et al., (1967) that both D and H responses could be due mainly to an increase in Cl permeability.

In the L cells, the synaptic action of L_{10} is simulated by ACh (Kunze & Brown, 1971). Moreover,

$$E_{ACh} = E_{IPSP} = E_{Cl} \qquad (4)$$

which means

1) the synapses for the "early IPSP" are electrically near the cell body (see discussion following equation (5))

2) a^i_{Cl} measurements are reliable in these cells and are not invalidated by some other anion to which the Cl ion exchanger may be sensitive. However,

$$E'_{ACh} = E_K \neq E'_{IPSP} \qquad (5)$$

This suggest that some of the synapses for the "late IPSP" are remote from the cell body. In such a case, $E'_{ACh} \neq E'_{IPSP}$, since the polarizing potential in the cell body required to reverse the "late IPSP" at the remote synapses located along the axon would decay electrotonically. Such differences have been similarly accounted for by others (Rall, 1962; Burke and Ginsborg, 1956; Calvin, 1969). However, it is also possible that either a transmitter other than ACh may be involved in synaptic action or that E_{ACh} is different for synaptic and extrasynaptic sites as has been recently shown in frog neuromuscular junction by Mallart and Feltz (1969). The fact that $E'_{ACh} = E_K$ means that the cell body also has ACh receptors that produce an increase in K permeability. Finally, the a^i_K measurements in these cells seem reliable and there is probably no interference from competing cations.

Limitations and possible errors in using ion selective microelectrodes.

1) The measurements are an artefact (Ling, 1969). Ling attributes the values obtained by Lev (1964) and Hinke (1959) to the damaged microenvironment about the ion electrode tips. We may estimate the time required for the ions to diffuse out of this microenvironment as follows. From Crank (1956) and Hinke (1973), the reflected diffusion from a layer of thickness (h) of a substance with initial concentration (C_o at t = 0) may be described as

$$\frac{C}{C_o} = \text{erf}\,\frac{h-x}{2\sqrt{Dt}} + \text{erf}\,\frac{h+x}{2\sqrt{Dt}} \qquad (6)$$

where C is the concentration at distance x, and D the diffusion coefficient. Taking D as roughly 1/10 its value in free solution makes it 10^{-6} cm^2 sec^{-1}, and making h about 1000 Å predicts that about 10^{-2} sec would be required before the activity near the electrode tip drops to 10% of its value. Since our measurements are always made over periods greater than 30 minutes (see above) following insertion of the electrode, the damaged microenvironment should be equilibrated with the remainder of the cytoplasm. Restrictions of ionic diffusion are unlikely in view of the results of Hodgkin and Keynes (1955) and Weidman (1963) concerning the intracellular diffusion coefficient of ^{42}K. Moreover, our data taken with those of Sato, et al., (1968) are strong evidence that K^+ is mainly free in the cytoplasm of the *Aplysia* neurons (see above).

2) The measurements in the vicinity of the ion selective microelectrode, even if true, may not be representative of the entire cell. Or in the case of *Aplysia* neurons, they are intranuclear and not cytoplasmic. We have made measurements in different parts of the same cell, including the trophospongium where there is no nucleus, and the measurements are identical. While it is true that the cell bodies of *Aplysia* neurons are largely occupied by the nucleus (Coggeshall, 1967), it is likely that the nucleus and cytoplasm are in equilibrium insofar as the major ions carrying currents are concerned. Thus Loewenstein and Kanno (1964) found transnuclear potentials of about OmV and a nuclear membrane resistivity of $15\Omega cm^2$, which is about 1.5×10^{-4} that of the cell membrane of *Aplysia* (Russell & Brown, 1972b).

The possibility of unstirred layers inside and outside the cell membrane (which has a number of infoldings) may exist. It is unlikely that these are important since the values for E_{ACh} and therefore E_i, the ionic equilibrium potential of the ion of interest are accurately reflected by the bulk phase ionic-activity measurements made inside and outside the neuronal membrane.

3) Although the ionic activities measured using ion microelectrodes are accurate, the intracellular ions are largely bound except for a small space with which they and the cell membrane are in equilibrium. It is into this space that the ion electrodes are inserted. This argument arises from the observation of Carpenter, Hovey, and Bak (1971) that the intracellular conductance of *Aplysia* nerve cell bodies is 10% that of sea water. They attribute this to a structuring of water which restricts the ability of the ions to carry electrical current. If this were the case it is difficult to explain the finding that multiple impalements of the same neuron with ion selective microelectrodes yield similar results.

4) The ionic activity measurements may include contributions from organic ions for which the electrodes are not selective. R_2 behaves like a Cl^- electrode when exposed to ACh (Russell & Brown, 1972a), and L cells behave like Cl^- electrodes for the early IPSP and K^+ electrodes for the late IPSP. Hence a comparison of the reversal potentials of these effects provides another estimate of the appropriate E_i's. The finding that these two different measurements yielded similar values for ionic activities supports the accuracy of the ion selective microelectrode measurements.

5) Leakage of ion exchanger material may damage the cells and invalidate the measurements. We checked this by measuring K^+ using NAS 11-18 glass having K_{NaK}'s varying over a range of 5:1, (Lev, 1964). The K^+ values were not different using either liquid ion-exchanger microelectrodes or glass microelectrodes. We have also made extensive comparisons of activity measurements with the ion exchanger microelectrodes and concentration measurements using flame photometry and have found them to be entirely consistent (Cornwall, et al., 1970; Chow, et al., 1970).

CONCLUSION

The evidence strongly favors the conclusion that measurements of the intracellular activities of K^+, Na^+ and Cl^- made with ion selective microelectrodes inserted into nerve cell bodies are accurate. The measurements are important because the appropriate ionic equilibria potentials driving ions across cell membranes can be calculated.

REFERENCES

Brown, A.M., J.L. Walker, Jr., and R.B. Sutton. 1970. Increased chloride conductance as the proximate cause of pH effects in Aplysia neurons. J. Gen. Physiol. 56: 559-582.

Brown, H.M., and A.M. Brown. 1972. Ionic basis of the photoresponse of Aplysia giant neurone: K^+ permeability increase. Science 178: 755-756.

Burke, W., and B.L. Ginsborg. 1956. The action of the neuromuscular transmitter on the slow fibre membrane. J. Physiol. 132: 599.

Calvin, W.H. 1969. Dendritic synapses and reversal potentials: theoretical implications of the view from the soma. Exp. Neurol. 24: 248.

Carpenter, D.O., M.M. Hovey, and A.F. Bak. 1971. Intracellular conductance of Aplysia neurons and squid axons as determined by a new technique. Intern. J. Neuroscience 2: 35-48.

Chiarandini, D. J., E. Stefani, and H. M. Gerschenfeld. 1967. Ionic mechanisms of cholinergic excitation in molluscan neurons. Science 156: 1957.

Chow, S.Y., D.L. Kunze, A.M. Brown, and A.M. Woodbury. 1970. Chloride and potassium activities in luminal fluid of turtle thyroid follicles as determined by ion selective ion-exchanger microelectrodes. Proc. Natl. Acad. Sci. 67 998-1004.

Coggeshall, R.E. 1967. A light and electron microscope study of the abdominal ganglion of Aplysia californica. J. Neurophysiol. 30: 1263-1287.

Cornwall, M.C., D.F. Peterson, D.L. Kunze, J.L. Walker, Jr., and A.M. Brown. 1970. Brain Research 23: 433-436.

Crane, J. In press. The mathematics of diffusion. Oxford University Press, London.

Eaton, D.C. 1972. Potassium ion accumulation near a pace-making cell of Aplysia. J. Physiol. 224: 421-440.

Frazier, W.T., E.R. Kandel, I. Kupferman, R. Waziri, and R.E. Coggeshall. 1967. Morphological and functional properties of identified neurons in the abdominal ganglion of Aplysia californica. J. Neurophysiol. 30: 1288.

Gerschenfeld, H., and L. Tauc. 1961. Pharmacological specificities of neurones in an elementary central nervous system. Nature 189: 924.

Hinke, J.A.M. 1959. Cytology and genetics - glass micro-electrodes for measuring intracellular activities of sodium and potassium Nature 184: 1257-1258.

Hodgkin, A.L., and R.D. Keynes. 1955. The potassium permeability of a giant nerve fibre. J. Physiol. 128: 61-88.

Kehoe, J.S., and P. Ascher. 1970. Re-evaluation of the synaptic activation of an electrogenic sodium pump. Nature 225: 280.

Kerkut, G.A., and R.W. Meech. 1966. The internal chloride concentration of H and D cells in the snail brain. Comp. Biochem. Physiol. 19: 819.

Kunze, D.L., and A.M. Brown. 1971. Internal potassium and chloride activities and the effects of acetylchloine on identifiable Aplysia neurons. Nature 229: 329-331.

Lev, A.A. 1964. Determination of activity and activity coefficients of potassium and sodium ions in frog muscle fibres. Nature 201: 1132-1134.

Ling, G., and F.W. Cope. 1969. Potassium ion: Is the bulk of intracellular K^+ adsorbed. Science 163: 1335-1336.

Lowenstein, W.R., and Y. Kanno. 1964. Studies on an epithelial (gland) cell junction. I. Modifications of surface membrane permeability. J. Cell Biology 22: 565-586.

Mallart, A., and A. Feltz. 1969. Mise en jeu de permeabilities ioniques distinctes au niveau des recepteurs synaptiques et extrasynaptiques des fibres musculaires striees. C.R. Acad. Sci. 268: 2724.

Orme, Frank. 1969. Liquid ion exchanger microelectrodes in Glass Microelectrodes. Ed. Marc Lavallee, Otto Schanne and Normand Hebert. John Wiley and Sons, New York.

Pinsker, H., and E.R. Kandel. 1969. Synaptic activation of an electrogenic sodium pump. Science 163: 931.

Rall, W. 1962. Theory of physiological properties of dendrites. Annals N. Y. Acad. Sci. 96: 1071.

Russell, J.M., and A.M. Brown. 1972a. Active transport of potassium by the giant neuron of the Aplysia abdominal ganglion. J. Gen. Physiol. 60: 519-533.

Russell, J.M., and A.M. Brown. 1972b. Active transport of chloride by the giant neuron of the Aplysia abdominal ganglion. J. Gen. Physiol. 60: 499-518.

Sato, M., G. Austin, H. Yai, and J. Maru haski. 1968. The ionic permeability changes during acetylcholine-induced responses of Aplysia ganglion cells. J. Gen. Physiol. 51: 321-345.

Sorokina, Z.O. 1966. Activity of sodium and potassium ions in giant neurones of molluscs. Fiziol. Zh. Kiev. 12: 776.

Thomas, R.C.1969. Membrane current and intracellular sodium changes in a snail neurone during extrusion of injected sodium. J. Physiol. 201: 495.

Walker, J. L., Jr. 1971. Liquid ion exchanger microelectrodes for Ca^{++}, Cl^- and K^+. In Ion Selective Microelectrodes. N.C. Hebert and R.H. Khuri, eds., Dekker, New York, in press.

Walker, J. L., Jr. and A.M. Brown. 1970. Unified account of the variable effects of CO_2 on nerve cells. Science 167: 1502.

DISCUSSION

Question: (Krnjevic)

1. Are the dimensions of the cells and the membrane fluxes large enough to allow appreciable ionic concentration gradients to develop ions to the cells?
2. Have you tried to block the chloride pump?
3. Have you tried using Copper to block?

Answer: (Brown)

1. The maximum time for diffusion is about 1 sec. This is calculated from $\lambda = \sqrt{Dt}$ where λ is distance and D is diffusion coefficient. Our readings were averaged over 5 mins so that distribution is probably homogeneous.
2. Yes - Ouabain partially blocks it. The effects of Ethacrynic acid & ameloride have been tried but we have not analyzed the data.
3. No.

Question: (Berman)
Have you any evidence for the presence of activity gradients for specific ionic species within the cell?

Answer: (Brown)
No. But this doesn't mean that they don't exist of course.

Question: (Durst)
1. Were the voltage-sensing electrodes and reference electrodes the same type?
2. How did you calculate the activity errors you quoted?
3. What are the problems associated with the development and fabrication of a calcium ion-selective microelectrode?

Answer:
1. Yes
2. I added the reading errors on each of 3 voltages and calculated the error from the calibration curve.
3. a) sensitivity - 10^{-6} to 10^{-8} M required
   b) selectivity - 10^{6} over monovalent cations
   10^{4} over divalent cations.

III. Kidney

USE OF POTASSIUM ION-EXCHANGER ELECTRODE FOR MICROANALYSIS

Fred S. Wright
Department of Physiology
Yale University School of Medicine

New Haven, Connecticut 06510

Numerous constituents of renal tubule fluid have been measured *in vitro* after collecting small volumes in micropipets. Some of these (hydrogen, sodium, and potassium ions) have also been measured *in vivo* by direct puncture of surface tubules with microelectrodes (1-6). Similarly the intracellular ionic composition of renal tubule cells has been estimated chemically in digests of renal tissue by several workers (7-9) and, recently, some results of direct measurements with potassium sensitive electrodes have been reported (6,9). Knowledge of cellular and luminal potassium concentrations in mammalian renal tubules, especially in the distal portion of the nephron, is crucial to an understanding of the mechanisms by which potassium is transported and electrical potentials are generated. In the course of attempting *in vivo* measurements of intracellular and intratubular potassium activity we have been confronted by the special problems attending measurements in the mammalian distal tubule. The presence of at least two unknown compartments in and across the epithelium with different sodium and potassium concentrations (10,11), different electrical potentials, and different total ionic strength (12) are difficulties that are compounded by the small size of mammalian cells--at most 10 microns in diameter. In addition to the problem of localization of the tips of the exploring K selective electrode and the reference electrode, being able to predict the behavior of the reference electrode is critical to evaluating changes in potassium activity. The tendency of potentials being registered by microelectrodes to change when the solution surrounding the pipet tip is diluted (13,14) is of concern in mammalian distal tubules because the luminal fluid is hypoosmotic with regard to plasma and the total ionic strength may range from about 50 to above 100mM. Potassium concentration in this fluid may be higher or lower than plasma concentration while sodium concentration ranges from about 1/5 to 1/2 of plasma values.

To gain a better understanding of how reference electrodes behave in this situation we have made a series of *in vivo* measurements using liquid ion-exchanger microelectrodes. Having had experience with emission photometry methods we know that although both the gas-oxygen microflame photometer (10) and the helium glow photometer are usually adequate to measure sodium and potassium (15,16), neither is ideal with respect to stability and reproducibility--especially with sample volumes less than one nanoliter. Therefore, in addition to examining reference electrode properties, we also wished to explore the feasibility of liquid ion-exchanger microelectrodes to measure potassium in nanoliter samples of renal tubule fluid. To do this we have compared measurements made with K selective electrodes with measurements of potassium ion in the same samples using a gas-oxygen microflame photometer.

The single barrel K selective electrodes for these experiments were made from Pyrex glass pulled with a tip diameter of about one micron. The tips were siliconized using a 1% solution of a chlorosilane product (SC-87, Pierce Chemical Co.) in toluene. Liquid ion-exchanger (Corning 477317) was introduced from the shaft of the pipet and pushed to the tip.

With regard to the reference electrode, we found it necessary to minimize any change in tip potential to attain the best sensitivity. When the reference electrode is connected to the ground side of the circuit, dilutional changes severely affect the K potential measured by the K selective electrode. An example is shown in Fig. 1 in which electrode potentials are plotted against the concentration of NaCl or KCl in the bathing solution. The top pair of lines are for two reference electrodes with the same tip size, both filled with 3 M LiAc, both measuring against a 3M KCl bridge with a diameter of 2 mm, but one with agar dissolved in the filling solution. I will say more about the agar electrodes in a moment. For now I want to call your attention to the increasingly more negative tip potential shown by the agar filled electrode. The slope in the region below 100 mM is about 20 mV per decade.

The pair of lines at the bottom of Fig. 1 connect data from a K selective electrode immersed with either of the reference electrodes in NaCl solutions ranging from 5 to 500 mM. The middle pair of lines are for the same K electrode and reference electrodes in KCl solutions ranging from 10 to 500 mM. The more positive potential recorded in the KCl solutions indicates a potassium over sodium selectivity of about 100:1. The measurements against the aqueous filled reference electrode have a slope of 56 mV per 10

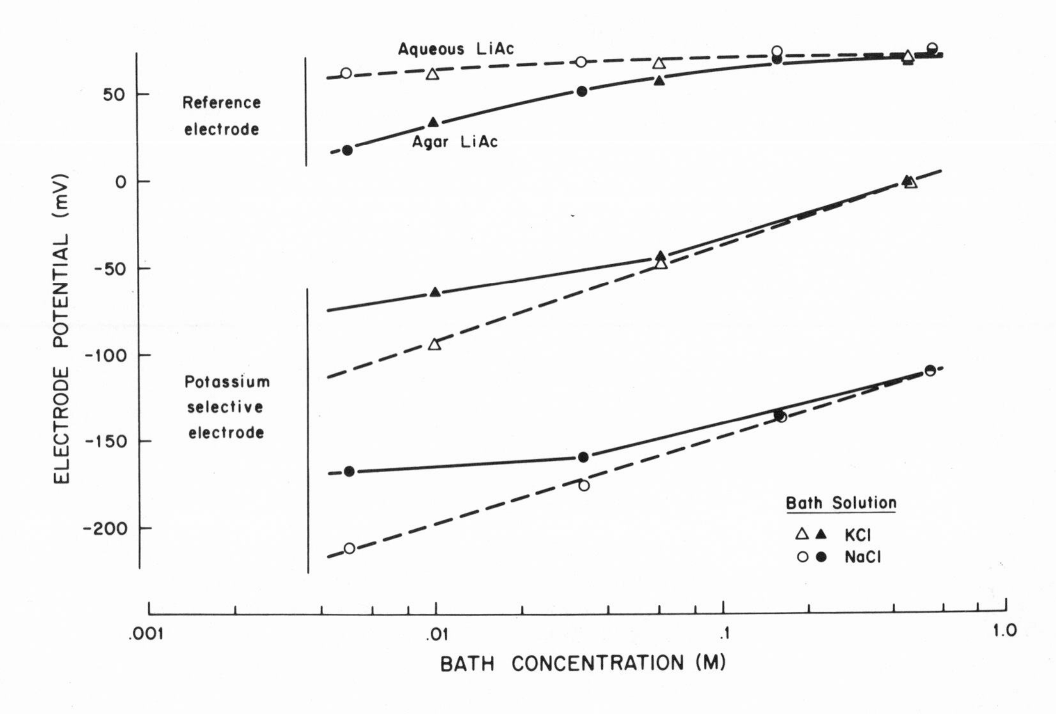

Fig. 1. Effect of reference electrode on potassium determination. Single measurements in NaCl or KCl with reference electrode (glass micropipet with 5 μ m tip) against a 3 M KCl agar bridge 2 mm in diameter, upper lines, and with liquid ion-exchanger electrode against the reference electrode, lower lines. The reference electrode was filled with a solution of 3 M lithium acetate with or without addition of agar 2 g/100 ml.

fold changes in potassium. Against the agar electrode, however, the slope is only about 35 mV. The changes in the reference electrode potential subtract from the simultaneous changes in the K electrode potential. The agar electrode thus exhibits partial cation selectivity but does not favor potassium over sodium. This change in slope response of the measuring system decreases the sensitivity near 10 mM K by about three times--a 1 mV change in potential corresponds to a 1.2 mM change in K concentration rather than a 0.4 mM change. We would like, therefore, to reduce or eliminate this change in reference potential.

One of the factors known to affect the magnitude of the change in tip potential is the tip diameter. In the extreme case the tip potential is eliminated when the tip is 40 μ m in diameter. (This is shown in Fig. 2).

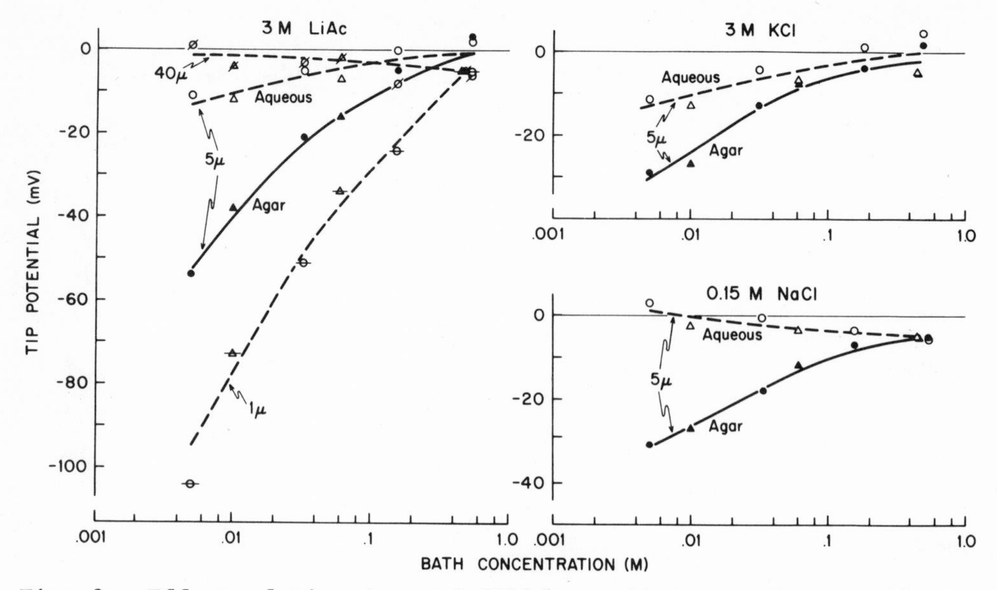

Fig. 2. Effect of tip size and filling medium on reference electrode tip potentials. Single measurements in NaCl or KCl with reference electrode against a 3 M KCl agar bridge. Left panel, lithium acetate filled electrodes with 1 μ m tip, broken tip (40 μ m), and 5 μ m tip (with and without agar). Right panels, pipets filled with KCl or NaCl with or without agar 2g/100 ml.

The lowest line in the left panel--the one designated 1 μ--traces changes in the tip potential, measured against a 3 M KCl bridge, of a LiAc filled microelectrode immersed in varying concentrations of NaCl and KCl. This pipet had a tip resistance of 20 megohm. Going from 150 to 50 mM the tip potential became 18 mV more negative. When the tip was broken off leaving a 40 μ m opening and a 2 megohm resistance, the reference potential--as shown by the top line in this panel--no longer declined in dilute solutions. A sharpened micropipet with a 5 μ m tip and a tip resistance of 3 megohm showed a small change in tip potential. Between 150 and 50 mM the tip potential became 2.5 mV more negative.

We, therefore, first used the pipet with the broken tip as a reference electrode with a K ion-exchanger electrode in drops placed under oil. This pipet proved to be unsatisfactory as a reference electrode, however, not because of changes in tip potential, but because, upon being inserted into a drop of a few nanoliters in volume, the sample disappeared into the pipet. Some barrier to free exchange between the sample and the filling solution is required to permit measurement in small samples.

To avoid the problem of aspiration of the samples, we added 2% agar to the fluid used to fill the reference pipet. As shown in Figs. 1 and 2, however, potentials measured with agar filled pipets were also found to change as the external solution is diluted. The middle lines in the left panel of Fig. 2 and the right panels show results obtained with 5 μ m pipets filled with either aqueous or agar solutions of 3 M LiAc, 3 M KCl bridge, 2 mm in diameter. Between 150 and 50 mM the change in the potentials of the agar electrodes was more negative than that of the aqueous filled electrodes by 7.5 mV for LiAc and NaCl, and 2.5 mV for KCl. The effect of these agar related changes on measurements with a K electrode would be reductions in sensitivity of 10% for agar-KCl, 25% with NaCl and 50% with LiAc. Since changes in the tip potential of the aqueous filled pipet with 5 μ m tips were small, we sought a way to exploit this advantage without paying the penalty of having a capillary that would aspirate the sample. It seemed that applying some small hydrostatic pressure to the solution filling the electrode might accomplish this aim.

A pipet holder that seals the rear opening of the pipet into a chamber that is closed except for a side arm is useful for this purpose. The holder we used is shown in Fig. 3. The gasket seals the pipet into the chamber and additional filling solution extends a variable distance above the level of the pipet. With a 5 μ m

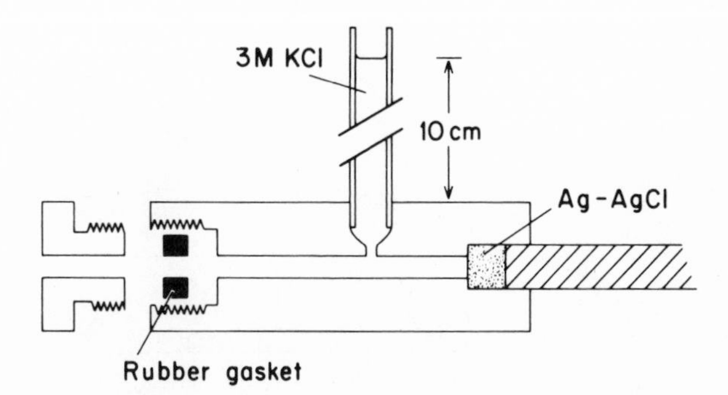

Fig. 3. Micropipet holder. Screw collar and gasket seal pipet in chamber; side arm permits a column of filling solution of variable height to exert pressure.

pipet in this holder and a fluid column of about 10 cm we found no movement of small samples into the pipet. If the pressure is high enough of course an outward flow of filling solution will occur. With 10 cm H_2O pressure the rate of outflow was less than 1/10 nanoliter per minute. It is a simple matter to reduce the pressure enough to eliminate this outflow. This type of pipet holder also appears to solve another difficulty we and others have encountered. This is the tendency for the liquid ion-exchanger in the tip of the K selective electrode to be displaced up into the pipet. Low pressure applied against the filling solution and ion-exchanger appears to prevent this displacement and thus prevents loss of responsiveness and selectivity.

The next step in characterizing this electrode system was to define the contribution of sodium in the test solution to the reading of the K electrode. Results of measurements in small drops of solutions with varying potassium and sodium concentration are shown in Fig. 4. The reference electrode for these measurements was connected to a 5 μ m pipet filled with 3 M LiAc. The reference potential of this pipet changes more than that of a pipet filled with .15 M NaCl, but we selected the lithium solution for our initial measurements because the selectivity of the K ion exchanger is about ten times less for lithium than for sodium. The lower line connecting the open circles shows the response in solutions of

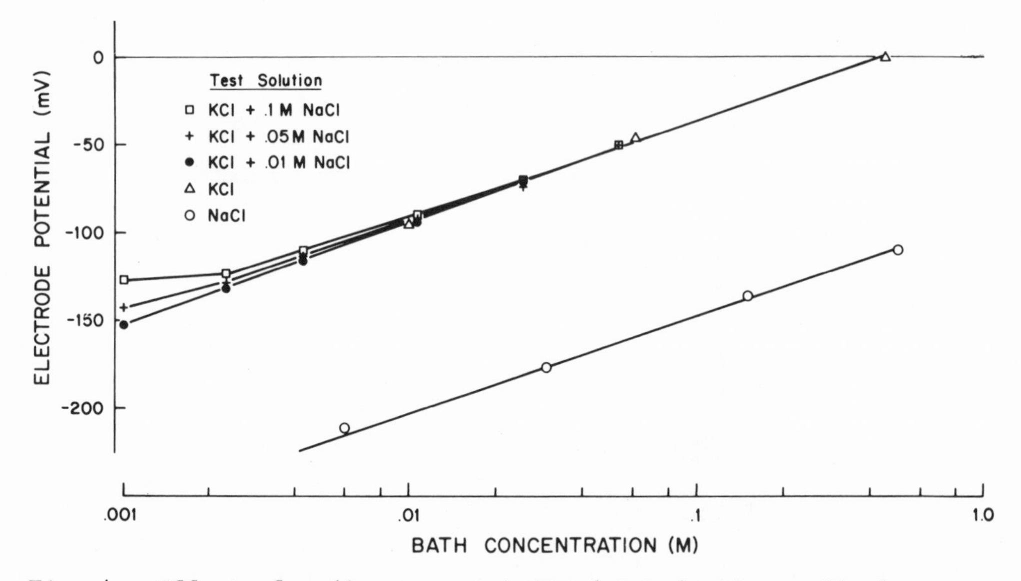

Fig. 4. Effect of sodium on potassium determination. Single measurements with ion-exchanger microelectrode in pure solutions and mixtures of NaCl and KCl.

NaCl. The open triangles above are from measurements in KCl alone. The slope of this line is 58 mV per decade and indicates a K:Na selectivity of about 100:1. The remaining points were obtained from measurements in drops placed under mineral oil. Data from samples with K concentrations ranging from 1 to 30 mM with a constant background of 10 mM Na did not deviate appreciably from the zero Na line. With 50 or 100 mM sodium in each sample there were significant deviations. For example, if a solution had a potassium concentration of 3 mM, and no correction for sodium was made, addition of 50 mM Na would give a potential indicating 3.4 mM K. To examine accuracy and reproducibility we compared measurements of the same solution by macro-flame photometer, micro-flame photometer and the potassium electrode. Fig. 5 shows results of these determinations. Analysis of microliter volumes by flame photometry showed the solution contained 5.9 mM KCl and 60 mM NaCl. Measurement of nanoliter volumes of this solution with the micro-flame photometer and the K electrode yielded mean values for potassium concentration of 5.6 and 4.8 mM. Single measurements in separate

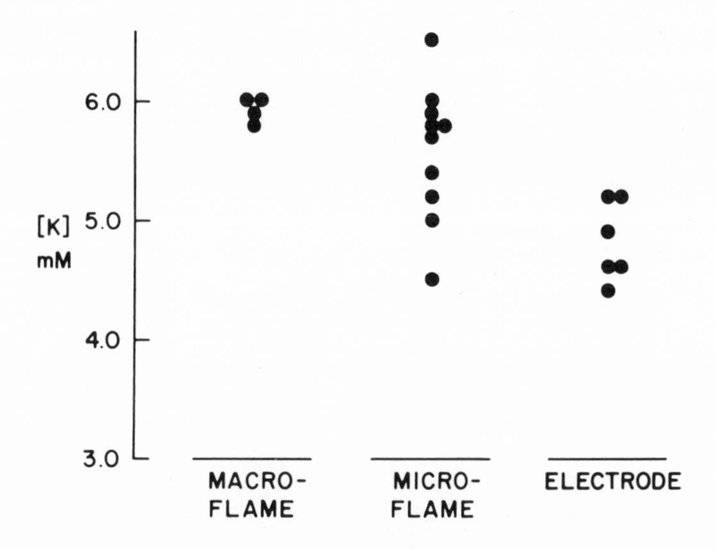

Fig. 5. Comparison of determinations of potassium concentration in a solution containing 60 mM NaCl using a commercial flame photometer, a micro-flame photometer and an ion-exchanger microelectrode.

drops with the K electrode varied over a range of 4 mV. This scatter was less than that with individual measurements using the microflame photometer. The differences between mean values do not seem to be due to the presence of sodium. The values are taken from the calibration line in the previous slide for solutions containing 50 mM Na. If sodium had been neglected the measured potential would correspond to a potassium concentration of 5.3. The K electrode measurements were made in drops ranging in size from 2 to 100 nl. There was no correlation between drop size and the variation in potential.

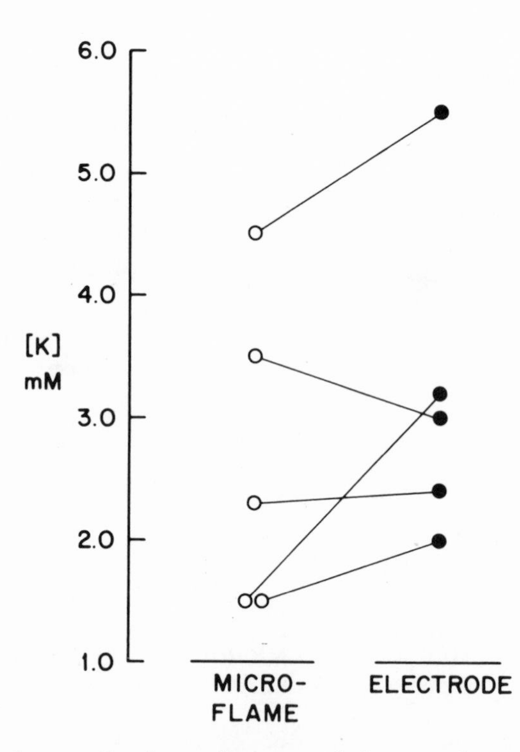

Fig. 6. Comparison of micro-flame and ion-exchanger electrode determinations of potassium concentration in five samples of fluid collected from distal tubules of rat kidney.

In another approach to determining the accuracy of the K electrode we compared measurement of five different samples of fluid obtained by micro-puncture from renal distal tubules of rats. The unknowns were read against the same set of calibrating standard solutions with the micro-flame photometer and with the potassium electrode. Fig. 6 shows the results of these measurements. Four of the measurements using the K electrode were from 0.1 to 1.7 mEq/L higher than measurements of the same sample with the micro-flame photometer. One sample read 0.5 mEq/L lower using the K electrode. Since in this series of determinations the K electrode result averaged .5 mEq higher than the flame photometer result, it appears that the lower mean on the previous slide is not due to a systematic tendency to underestimate. The higher readings shown in this figure are not due to a contribution of sodium. The sodium concentration determined simultaneously in the micro-flame measurements ranged from 27 to 33 mM in the four overestimated samples and was 55 mM in the sample which the K electrode read lower. Thus, there is no suggestion in these measurements of systematic differences due to variations in sodium concentration. A possible source of these differences is interference by ammonium ion. The potassium selectivity of the Corning 477317 exchanger is only 5:1 over ammonium (17). Since the concentration of ammonium in distal fluid may equal that of potassium, an overestimate of up to 20% might be attributed to ammonium. This was approximately the degree to which K was overestimated on the average in samples of tubule fluid. Simply the absence of ammonium, however, does not explain the degree to which K was underestimated in the solution containing only sodium and potassium chloride.

The main advantages of the K selective electrode as a microanalytical tool appear to be that it does not consume the sample, less time is needed to make the measurements and the equipment is less expensive for initial cost. Further testing may show that superior accuracy can be obtained with the electrode in very small volumes down to about 0.1 nanoliter.

The K electrode does not give a simultaneous measurement of sodium concentration, and, although the range of sodium concentration in these samples does not seem to interfere with the measurements, the presence of ammonium may need to be taken into account.

REFERENCES

1. Rector, F.C., Carter, N.W. and Seldin, D.W. The Mechanism of Bicarbonate Reabsorption in the Proximal and Distal Tubules of the Kidney. J. Clin. Invest. 44:278-290. 1965.

2. Vieira, F.L. and G. Malnic. Hydrogen Ion Secretion by Rat Renal Cortical Tubules as Studied by an Antimony Microelectrode. Am. J. Physiol. 214:710-718, 1968.

3. Khuri, R.N., Goldstein, D.A., Maude, D.L., Edmonds, C. and Solomon, A.K. Single Proximal Tubules of Necturus Kidney. VIII: Na and K determinations by Glass Electrodes. Am. J. Physiol. 204:743-748, 1963.

4. Khuri, R.N., Flanigan, W.J., and Oken, D.E. Potassium in Proximal Tubule Fluid of Rats and Necturus measured with Glass Electrodes. J. Appl. Physiol. 21:1568-1572, 1966.

5. Khuri, R.N. Cation and Hydrogen Microelectrodes in Single Nephrons. In Glass Microelectrodes, M. Lavallee, O.F. Schanne, and N.C. Hebert, eds. John Wiley, N.Y., 1969. pp272-298.

6. Khuri, R.N. Intracellular Potassium in Cells of the Distal Tubule. Pflugers Arch. 335:297-308, 1972.

7. Burg, M., Grantham, J.J., Abramow, M. and Orloff, J. Preparation and Study of Fragments of Single Rabbit Nephrons. Am. J. Physiol. 210:1293-1298, 1966.

8. Abramow, M., Burg, M.B., Orloff, J. Chloride flux in Rabbit Kidney Tubules In Vitro. Am. J. Physiol. 213:1249-1253, 1967.

9. Khuri, R.N., Hajjar, J.J., Agultan, S., Bogharian, K., Kalloghian, A. and Bizri, H. Intracellular Potassium in Cells of the Proximal Tubule of Necturus maculosus. Pflugers Arch. 338:73-80, 1972.

10. Malnic, G., Klose, R.M. and Giebisch, G. Micropuncture Study of Renal Potassium Excretion in the Rat. Am. J. Physiol. 206:674-686, 1964.

11. Malnic, G., Klose, R.M. and Giebisch, G. Micropuncture Study of Distal Tubular Potassium and Sodium Transport in Rat Nephron. Am. J. Physiol. 211:529-547, 1966.

12. Wright, F.S. Increasing Magnitude of Electrical Potential along the Renal Distal Tubule. Am. J. Physiol. 220:624-638, 1971.

13. Agin, D.P. Electrochemical Properties of Glass Microelectrodes. In. Glass Microelectrodes. M. Lavallee, O.F. Schanne and N.C. Hebert. John Wiley & Sons, N.Y., 1969. pp62-75.

14. Lavallee, M. and Szabo, G. The Effect of Glass Surface conductivity Phenomena on the tip Potential of Micropipette Electrodes. In Glass Microelectrodes. M. Lavallee, O.F. Schanne and N.C. Hebert. John Wiley & Sons, N.Y. 1969, pp95-123.

15. Vurek, G.G. and Bowman, R.L., Helium-glow Photometer for Picomole Analysis of Alkali Metals. Science. 149:448-450 1965.

16. Vurek, G.G. Emission Photometry of Picomoles of Calcium, Magnesium and other Metals. Anal. Chem. 39:1599-1601, 1967.

17. Wright, F.S. and McDougal, W.S., Potassium-specific Ion-exchanger Microelectrodes to Measure K^+ activity in the Renal Distal Tubule. Yale J. Biol. Med. 45:373-383, 1972.

DISCUSSION

Question: (Brown)

What is the effect of 10 cm hydrostatic pressure on time constant, DC voltage and selectivity?

Answer: (Wright)

Applying a small pressure behind the K ion-exchanger seems to prevent displacement of the exchanger back away from the tip. Because of this it prevents the formation of an aqueous chamber in the tip and this prevents loss of sensitivity and response. We have not systematically studied the relation between different pressures and the electrode potential.

Question: (Joel Brown)

What are the origins and measurement conditions for the selectivity data shown?

Answer: (Wright)

They were collected from several sources: data from Corning, from published results of Wise et al, from published work by Walker and from our own measurements all using the Corning exchanger 477317.

Question: (Krnjevic)

Does the application of pressure result in a detectable change in tip resistance?

Answer: (Wright)

We have not determined the effect of small pressures on the resistance of the K electrode. One might expect that this resistance which is of the order of 10^{10} ohms would not be affected.

Questions: (Morris)

1. Comment after Dr. Hebert's question about 2 kinds of Corning K^+ ion-exchangers. There were two - one (?477162 no.) is an older one - mentioned in Khuri's early papers . Only one has been available for the last 1 1/2 years - 477317 -- which, presumably, has greater selectivity than the first one.

2. Were the calibrations you described -- with different concentrations of K^+ and Na^+- made with the LiAc, 5 μ tip reference electrode?

Answers: (Wright)

1. All of these data were obtained using the newer exchanger material 477317 (contains tetraphenylborate).

2. Yes, with a column of 3 M LiAc a few cm high behind the pipet.

KINETIC ANALYSIS OF RENAL TUBULAR ACIDIFICATION BY ANTIMONY MICROELECTRODES

Gerhard Malnic, M. Mello Aires and A.C. Cassola

Dept. of Physiology, Instituto de Ciencias Biomedicas
University Sao Paulo
Sao Paulo, Brasil

INTRODUCTION

Microelectrodes for the determination of renal tubular fluid pH have been used since the work of Montgomery and Pierce (14) on the amphibian nephron. These authors as well as others (2,3) used the quinhydrone electrode, where tubular fluid is aspirated into a micropipette and measured outside the tubular lumen, in general after equilibration with a known pCO_2. pH sensitive glass microelectrodes were used by Rector and Carter (15). These authors used spear-type electrodes, which are externally insulated but for the very tip, permitting measurements within the tubular lumen. Microelectrodes where the sample is drawn up into a pH-sensitive glass capillary have also been used (7,19).

The use of antimony microelectrodes which are metal/metal oxide electrodes, is shown in Fig. 1a. They consist of antimony-filled glass capillaries, which are drawn out in order to obtain a fine bevelled tip, which is pH-sensitive and can be introduced into the tubular lumen. These electrodes have also been used to measure titratable acidity and ammonia in an "in vitro" system, according to Solomon et al (16) and Karlmark (6). In this paper we will restrict ourselves to the discussion of some aspects of tubular acidification which can be studied due to the rapid response of the antimony electrode system to pH changes, permitting the kinetic study of the tubular acidification mechanisms.

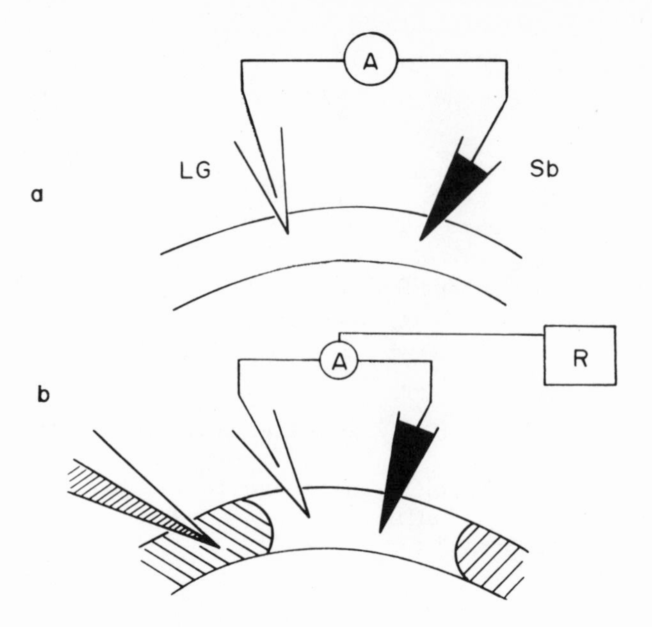

Fig. 1. Schematic diagram of the use of antimony microelectrode for tubular pH measurements. a, pH determination during free flow. b, continuous recording of tubular pH during stopped flow microperfusion.

METHOD

The preparation and general characteristics of the antimony microelectrode system have been described in detail previously (13,20). A schematic diagram showing the use of this system for the recording of continuous changes in luminal pH during microperfusion experiments is given in Fig. 1b. The tubule is perfused with buffered solutions which are isolated by Sudan-black colored castor oil columns and brought into contact with the antimony microelectrode and its reference electrode, which can be a Ling-Gerard type microelectrode. When no major potential difference (PD) across the measured nephron segment is expected, the potential of the Sb electrode can be measured against an extracellular reference. The measured potential difference is recorded continuously, as shown in Fig. 2. In the upper drawing of this Figure, the rapid change of the reading after the start of perfusion with alkaline buffer is seen. It approaches the new equilibrium level with a half-time of a few tenths of a second, faster than the subsequent physiological pH changes. After blocking the fluid column, a progressive acidification is noted, rather fast initially, and

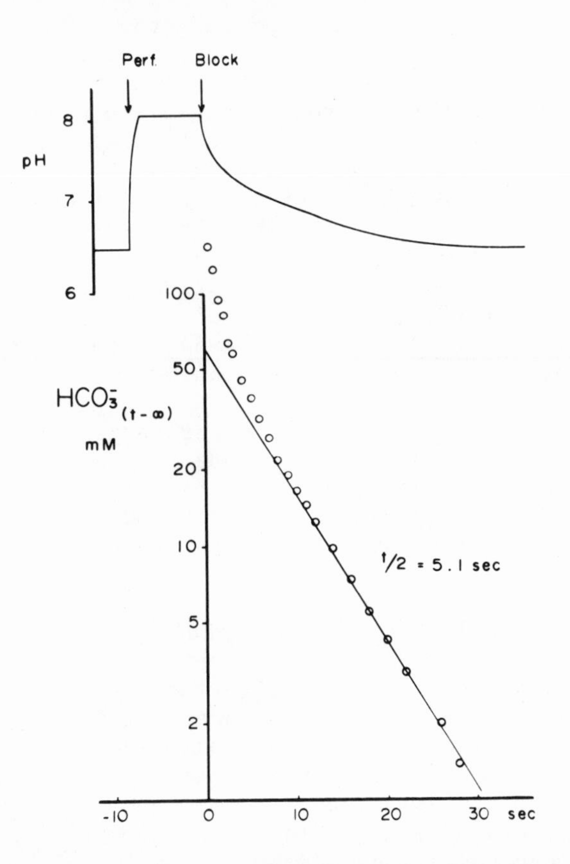

Fig. 2. Acidification of bicarbonate droplet in proximal tubular lumen. Upper graph, changes in luminal pH during perfusion with 100 mM bicarbonate and after blocking the tubule with oil. Bicarbonate concentrations calculated from this curve are given in the lower graph. Half time of bicarbonate reabsorption is obtained from slope of exponential line.

gradually approaching the luminal steady-state pH level. From such curves instantaneous bicarbonate concentrations can be calculated when the perfused fluid is preequilibrated with a physiological pCO_2, or after equilibration within the tubular lumen. Such values are shown on the lower graph of Fig. 2. When air-equilibrated solutions are used, the first part of the obtained curve can be used to evaluate the rate of equilibration of the injected fluid with the animal's pCO_2. Carbonic acid concentrations can be estimated from the measured pH and from bicarbonate concentrations extrapolated back to time zero from the slower exponential. The half-time of equilibration of both the fast CO_2-line and the slower bicarbonate reabsorption line can be obtained from such graphs (10).

A similar approach can be used for the study of the acidification of alkaline phosphate solutions. On the basis of pH curves obtained during such perfusions, acidification graphs are obtained plotting the difference between the steady-state acid phosphate concentrations and those obtained at time t, against time in seconds, on a semilog scale. In this case, a single exponential is obtained from the start of the recording in spite of the use of air-equilibrated solutions. In phosphate perfusions performed during different experimental conditions, it was noted that in alkalotic rats the slope of the acidification line was similar to that obtained in controls, but the maximal acid phosphate gradients were considerably lower. On the other hand, after acetazolamide administration both the slope and the gradient were decreased (12).

Other variations of the described experimental set-up have been used. Peritubular capillaries can be perfused simultaneously with the luminal procedure for the evaluation of acidification rates. In this way it is possible to study the influence of alterations of peritubular parameters like pH and pCO_2 on the acidification process.

CRITIQUE OF THE METHOD

Beside some of the criticism that pertains to microperfusion experiments in general, a few points concerning the kinetic measurements with the antimony electrode system should be considered more specifically. One of them concerns the behavior of the electrode in bicarbonate solutions, where it has been claimed that more acid readings are obtained than those theoretically calculated from bicarbonate concentrations and pCO_2 as well as those read by glass pH electrodes (5). We have observed such effects occasionally in our experiments. A possible source of error could be the generation of junction or tip potentials at the reference microelectrode due to differences in ionic composition of standard buffers and samples. These were minimized using low resistance (about 3 megohm), freshly prepared glass microelectrodes, and 2.5 M KCl / 0.5 M KNO_3 solution to fill these electrodes (15). With these precautions, in a series of 66 measurements of 25 mM bicarbonate droplets equilibrated with 5% CO_2, performed as a control during both "in vivo" and "in vitro" experiments, 59% were within 0.1 pH unit and 86% were within 0.2 pH units of the glass electrode value, while from the remaining 9 measurements, 7 read a more acid and 2 a more alkaline pH. A mean difference of - 0.065 $\pm$ 0.018 units between antimony and glass electrodes was observed for all values that were significantly different from zero.

Another criticism that can be raised is the interference of variations in transtubular PD during the perfusions. This would affect especially the early phases of perfusion, as well as experiments where the pH is not read differentially between the antimony and a micro reference electrode in the tubular lumen, but against a remote extracellular electrode. Experiments where the trans-tubular PD is read by means of Ling-Gerard microelectrodes in the proximal lumen showed that the observed PDs change by less than 10 mV when these segments are perfused with bicarbonate or phosphate solutions, which represents a relatively minor error since PD changes due to pH alterations correspond to about 2 pH units, that is, about 100 mV. In distal tubules, on the other hand, differential readings against luminal reference microelectrodes are more important. In this case, double barrelled microelectrodes with one Sb and one reference barrel are of value. It should always be kept in mind that the segment of tubular epithelium which is analysed with respect to its acid secretory capacity is also punctured by the electrodes, a procedure potentially prone to the production of leaks. It is, therefore, important to use thin microelectrodes. We believe, however, that with some experience, curves deformed by leaks across the epithelium or around the tubular oil block can be recognized and eliminated. The finding of significantly more acid steady-state luminal pH levels than those of extracellular fluid, however, indicates that such leaks should not play an important part in our experimental procedure.

PHYSIOLOGICAL INTERPRETATION OF KINETIC DATA

1. CO_2 Equilibration Kinetics

When tubular segments are perfused with air-preequilibrated solutions, there occurs an initial phase of rapid pH changes that can be attributed to the equilibration of the fluid column with the animal's CO_2 tension, since it disappears when CO_2 pre-equilibrated solutions are perfused. Based on simple two-compartment kinetics, half-times for this equilibration have been calculated. Since relatively slow half-times (0.5 to 1.5 sec) were obtained, they were related to a diffusion delay of CO_2 across the tubular epithelium and not to hydration of this gas in the fluid column, which can proceed more rapidly even without carbonic anhydrase catalyzation (10). This reasoning is based on the existence of a pure CO_2 / HCO_3^- / water system within the tubular lumen. The diffusion of other buffers into the fluid column could consume carbonic acid and thus delay the equilibration of luminal and peritubular carbonic acid levels. Therefore, only those perfusions where such diffusion is negligible would lead to equilibration values limited only by CO_2 diffusion into the lumen. It appears, therefore, that the values obtained by this method represent an upper limit for the magnitude of diffusion velocity of CO_2

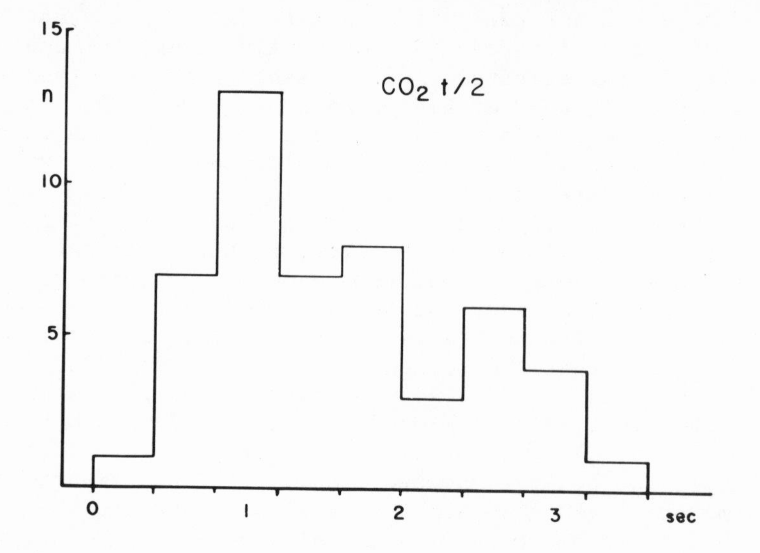

Fig. 3. Frequency distribution of half-times for CO_2 equilibration during perfusion of proximal tubules with air equilibrated solutions. Abscissa, half times in seconds. Ordinate, frequency of observations.

into the tubular lumen. In Fig. 3, it can be noted that there occurs a considerable dispersion of the half-times for CO_2 equilibration in bicarbonate droplets perfused in control rats. The lower limit of these equilibration half-times, of the order of 0.5 sec, may be a better estimate of CO_2 permeability of the tubular wall than the mean value. These values are of importance, since they can be used to evaluate CO_2 gradients building up during bicarbonate reabsorption from the tubular lumen. Such gradients contribute to the establishment of a disequilibrium pH, which was observed under different experimental conditions, and which has been related to disequilibrium levels of H_2CO_3, and more recently also of CO_2 (5, 8). This is clarified upon observation of the Henderson-Hasselbalch equation:

$$pH = pK + \log (HCO_3^-) / (H_2CO_3 + CO_2)$$

The acid term, which will be increased in the tubular lumen when a disequilibrium pH is found, depends on the two described factors, carbonic acid and CO_2.

The following calculations lead to an evaluation of transtubular CO_2 gradients. The loss of CO_2 from the tubular lumen is given by:

$$-\frac{dCO_2}{dt} = \dot{V}CO_2 = k_{CO_2} \cdot (\Delta CO_2)$$

where k_{CO_2} is a rate coefficient obtained from the half-time discussed above, and ΔCO_2 is the CO_2 concentration gradient across the tubular wall. On the other hand, CO_2 generation within the lumen is given by:

$$dCO_2 = \dot{V}'CO_2 = k_{HCO_3^-} (HCO_3^- t - HCO_3^- oo)$$

where $k_{HCO_3^-}$ is the rate constant for bicarbonate reabsorption due to H ion secretion, which is a reaction generating CO_2. The calculation of this constant will be described later; it is smaller than the overall coefficient obtained from experiments involving perfusion with bicarbonate solutions, and estimated by the rate of acidification of phosphate buffers. At steady-state, CO_2 loss and gain should be equal:

$$\dot{V}_{CO_2} = \dot{V}'_{CO_2}; \quad \text{thus,}$$

$$\Delta CO_2 = k_{HCO_3^-} (HCO_3^- t - HCO_3^- oo) / k_{CO_2}$$

According to this relation, the CO_2 gradient expected across the tubular lumen can be evaluated, when the involved rate coefficients are known, at any given luminal bicarbonate concentration. Such calculations are given in Table I for different bicarbonate levels,

TABLE I

Calculation of Transtubular CO_2 Gradients in Proximal Tubule.

Exp.	$(HCO_3^-)t$	$(HCO_3^-)oo$	$k_{HCO_3^-}$	k_{CO_2}	$\Delta(CO_2)$	Dis. pH
C	25 mM	4 mM	0.093	0.69 s^{-1}	2.83 mM	0.52
	25	4	0.093	1.39	1.41	0.34
	8	4	0.093	0.69	0.54	0.16
	8	4	0.093	1.39	0.27	0.08
D	20	7	0.053	0.18	3.82	0.55
	20	7	0.053	0.35	1.97	0.37

C: Control; D: Diamox.

and assuming k_{CO2} equal to 0.69 sec^{-1} (corresponding to a t/2 of 1 sec) or to 1.39 (corresponding to the lower limit of the observed half-time, that is, t/2 = 0.5 sec). Disequilibrium pH values obtained in these situations are also given. Data obtained in acetazolamide infused rats show disequilibrium values that are not much higher than those found in control rats at the same bicarbonate levels, since on one hand CO_2 equilibration is slow, but the rate of bicarbonate reabsorption is also reduced. These data indicate that observed disequilibrium pH values may be due not only to higher steady state carbonic acid concentrations, but also to transtubular P_{CO2} gradients; the latter are small in the middle third of the proximal tubule, but may be important in segments with higher bicarbonate concentrations. This could explain the disequilibrium pH that is found in proximal tubule of rats with respiratory acidosis, where higher bicarbonate concentrations are found in the accessible portions of the proximal tubule in comparison with controls (11). Obviously, it would be of great interest to evaluate this possibility in different experimental conditions by direct determination of tubular pCO_2 values.

2. H Ion Secretion vs. HCO_3^- Reabsorption

Bicarbonate ions can be reabsorbed by two mechanisms: transport of this ion as such across the tubular epithelium, or H ion secretion into the tubular lumen with ensuing decomposition of bicarbonate into CO_2 and water. The end results of these processes are equivalent, making their distinction a difficult problem, which has led to considerable controversy. I would like to discuss some aspects of the present method which could contribute to the solution of this problem. When the pH of a bicarbonate droplet is measured, the decrease in bicarbonate concentrations with time can be evaluated assuming a constant p_{CO2}. The rate of this decrease is proportional to the luminal bicarbonate concentration. However, all processes leading in parallel to a reduction in bicarbonate concentration toward the same steady-state level will participate in the measured acidification, indicating that the overall rate measured by this method may be the sum of several processes. The rate of disappearance of bicarbonate or any other luminal base is due to a sum of factors, each of them proportional to the same luminal concentration:

$$dHCO_3^- / dt = k'(HCO_3^-) + k''(HCO_3^-) = k'''(HCO_3^-)$$

where k' and k'' are individual rate coefficients, and k''' is the pooled coefficient.

Therefore, the rate coefficients related to these processes can be pooled into one composite rate coefficient, which is the experimentally measured value. One of the discussed factors appears to be quite certainly H ion secretion, as suggested by the occurrence of a disequilibrium pH under several experimental conditions, which is due to high luminal carbonic acid or carbon dioxide levels, both caused by the reaction between luminal bicarbonate and secreted H ion. The occurrence of this process, however, cannot rule out the existence of other mechanisms leading to reduction of bicarbonate concentration, like diffusion or transport of this ion out of the lumen. Specifically in the described microperfusion experiments, dilution by flow of NaCl and water into the tubular lumen when a chloride-free solution is perfused, is to be expected. This is demonstrated by measurements of volume variation with time, where upon perfusion with an isotonic sodium phosphate solution an increase in the volume of the fluid column is seen during the first 30 seconds. Only after some 40 sec does its volume starts to decrease, indicating that during the first phases of perfusion, water and salt flow into the column. The monitoring of chloride concentrations during this period shows a rapid increase in these concentrations, which tend to stabilize after 1 to 2 minutes (9). The equilibration rates of chloride are equivalent to anion disappearance rates, and could be used as a measure of buffer anion dilution by fluid inflow, a possible component of the acidification process of bicarbonate columns. It should be noted that they also include exchange of these anions against chloride, as well as any other mechanism of anion loss from the tubular lumen, including a possible component of HCl secretion in the case of bicarbonate.

On the other hand, acidification of phosphate buffer, assuming absence of loss of alkaline or acid salt by a specific transport mechanism, can be shown to evaluate H ion secretion. Actually, since there is evidence that there may be greater permeability to the acid salt which includes a monovalent anion, while the alkaline salt includes phosphate in the form of a divalent anion, such rates may be underestimates of H ion secretion (1). In Fig. 4, a comparison between bicarbonate and phosphate acidification rate coefficients is made based on this view. This comparison must be made based on the respective rate coefficients, since acidification rates depend on the product of such coefficients with the driving force for acidification, that is, the difference between initial and final buffer base concentration in the tubular lumen. For a valid comparison, these driving forces have to be equal. It can be noted that both in control and in acetazolamide infused rats the sum of the phosphate and the dilution coefficient approximates that related to overall bicarbonate reabsorption. Thus, bicarbonate reabsorption could be broken down into two main components, one of them being H ion secretion, and the other related to processes like dilution and transport of bicarbonate out of the lumen, as evaluated by chloride inflow into non-buffer anion solutions like sulfate. According to this estimate, at least 55% of the bicarbonate reabsorbed under the described experimental conditions is transferred via the H ion secretory system, while the remainder may be attributed to the above described processes measured by means of estimation of

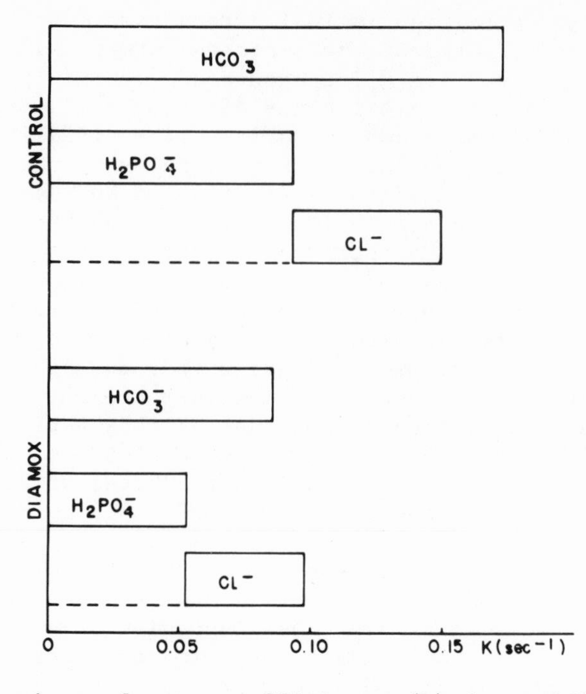

Fig. 4. Comparison of rate coefficients (k) for acidification during perfusion with bicarbonate (HCO_3^-) and phosphate ($H_2PO_4^-$) buffers, and for chloride equilibration (Cl^-), in proximal tubules of control and diamox infused rats.

chloride entry into the lumen. Bicarbonate reabsorption as such would be included in this estimate. The occurrence and magnitude of active bicarbonate reabsorption cannot be evaluated on the basis of the present data. However, considering the likelihood of significant dilution, considerably less than the remaining 45% should be attributable to possible active reabsorptive transport. Passive outward diffusion of bicarbonate should also not be a major component, since half-times obtained with 100 and 25 mM perfusions are of similar magnitude.

3. Mechanism of H Ion Secretion

Several models can be constructed in order to describe more formally the H ion secretory system. A conventional description would be based on a pump leak system, with a few simplifying assumptions in order to make a treatment based on the available data more feasible. Thus, a two-compartment system, based on a suggestion by Curran, could be used. Compartment 1 would be the tubular lumen, and compartment 2 an extraluminal space, for instance the tubular cell, visualized as the source of H ions for secretion. The basic equations for this system are the following:

$$dS_1/dt = - k_{12}S_1 + k_{21}S_2$$

which upon integration and introduction of the appropriate boundary conditions gives:

$$S_1 = S_{1\,oo} + (S_{1o} - S_{1\,oo})\ e^{-k_{12}t}$$

Here, S_1 represents the luminal acid phosphate concentration, which is zero at time zero ($S_{1o} = 0$), since only alkaline phosphate is perfused. It is, furthermore, assumed that the flow of H ions into the lumen, represented by $k_{21}\ S_2$, is maintained constant during the perfusion period, which implies that cellular pH is not changed during changes of luminal pH and buffer content. The variable term would be the backflow of H ions from lumen to extraluminal space, driven by the increasing luminal H ion secretion in the opposite direction. According to this interpretation, the rate coefficient k_{12}, which is obtained from the experimental data, is related to passive outflow of H ions from the tubular lumen, and therefore is a measure of the passive permeability of the tubular epithelium to these ions. This can be better visualised by the application of the Hodgkin-Katz equation (4), introducing the value for H ion flow, which leads to a relation between permability and the k_{12} coefficient.

$$J_H\ net = k\ (As - A)\ .\ r/2$$

where J_H is transepithelial H ion flow, A is the instantaneous acid concentration, As is this concentration at steady-state, and r is the tubular radius.

$$\text{In steady-state, } J_H\ in = J_H\ out$$

$$\text{At } t = 0,\ J_H\ in = k\ As.\ r/2 = J_H\ out \text{ at } t = oo$$

The permeability to H ions is given by:

$$P_H = \frac{J_H\ out}{C_{H\ in}} \cdot \frac{e^{EF/RT} - 1}{EF/RT}$$

where $C_{H\ in}$ is the luminal H ion concentration, in the present case equal to As. Substituting the indicated values,

$$P_H = \frac{k.\ As.\ r}{As.\ 2} \cdot \frac{e^{EF/RT} - 1}{EF/RT} = \frac{k\ .\ r}{2} \cdot \frac{e^{EF/RT} - 1}{EF/RT}$$

$$\text{If} \quad E = 0,\ P_H = \frac{k\ .\ r}{2} \text{ cm/sec}$$

In the absence of a transtubular PD, permeability and the coefficient k are equivalent but for a geometrical term. It is noted, also, that this coefficient is not directly related to permeability if the measurement is made in the presence of a transtubular PD, where it will have to be corrected according to the discussed equation. The application of this analysis to phosphate perfusion experiments is exemplified in Table 2. Results obtained

TABLE 2

H Ion Fluxes During Phosphate Buffer Perfusion

		S_{1oo} mM/1	k_{12} sec^{-1}	$k_{21}S_2$ mM/1.sec
Control	Prox	69.8	0.094	6.55
	E. Dist	40.2	0.097	3.90
	LD	53.3	0.081	4.32
5% $NaHCO_3^-$	P	36.6	0.101	3.70
	ED	16.3	0.082	1.34
	LD	20.9	0.095	1.99
Diamox	P	33.3	0.053	1.75
	ED	35.4	0.024	0.85
	LD	35.7	0.038	1.35

according to the discussed equations were calculated for control, acutely alkalotic and Diamox infused rats. The first column represents steady-state acid phosphate levels; the second, k_{12} rate constants calculated from the half-times obtained from the experimental graphs. $k_{12}\ S_2$ is the maximal unidirectional H ion secretory flux measured at t = 0, when backflux is zero. It can be noted that the k_{12} constants are of the same order of magnitude in proximal and distal tubules both for control and acutely alkalotic animals. In the latter, however, the steady-state level of acid salt is considerably lower, leading to a markedly decreased secretory rate, in the absence of permeability changes. In Diamox infused rats both permeability and steady state levels are decreased, leading to even lower secretory rates. As can be noted, this approach leads to a rational interpretation of the experimental data.

The pump-leak model, however, is not the only model that could be applied to tubular H ion secretion. For turtle bladder, Steinmetz et al (18) brought forward evidence in favor of a low permeability for H ions, the secretion of which is possibly being effected by an active transport mechanism whose activity is dependent on the gradient against which it pumps H ions. Thus, the pumping rate would be high when transport proceeds into an alkaline mucosal solution, and low when this solution is acid. In this case, the previously discussed flux equation should be altered. In the limiting case, where a linear relation between the rate of acidification and luminal acid or base concentration could be expected, permeability to H ions should be so low as to make it possible to neglect k_{12}, whereupon the secretory rate would be proportional to k_{21} S_2, the pumping rate. The latter could vary with luminal base level due to variation of S_2 or k_{21}. In this case, the measured rate coefficient would be an evaluation of the pump activity.

Some recently obtained results appear to be pertinent to this discussion. We have perfused proximal and distal tubules with 0.1 M phosphate solutions at an initial pH of 5.5, and recorded the alkalinization of these columns toward the steady state level. This is also an exponential process, and linear graphs relating the log of the difference between acid concentration at time t and steady state concentration with time are obtained. According to the previously given pump-leak model, the slopes of these lines should not be different from acidification lines, since they both depend essentially on the permeability of the epithelium to outflow of H ions from the tubular lumen. In our series, the difference between the mean values of the rate constants obtained in the proximal tubule during alkaline and acid perfusions was small, and of borderline significance, in view of the inherent uncertainties of the method, and we believe still compatible with the pump-leak model. In the distal tubule, however, these differences are quite considerable. The alkalinization line has a much greater half-time than that corresponding to acidification. The mean values of these half-times in proximal and distal (early) tubules are summarized in Table 3. It can be clearly seen that distal tubular half-times of alkalinization are considerably higher than acidification half-times, while in the proximal tubule the difference is minor. This observation suggests that in the distal tubule, especially in its early segment, a behavior similar to that of the turtle bladder may be found; that is, a low passive H ion permeability, as measured by the perfusions with phosphate at pH 5.5, may coexist with an H ion secretory pump whose activity is governed by the transtubular pH gradient, as evaluated by the acidification half-time. In the experiments de-

TABLE 3

Half-times of pH equilibration in cortical tubules of the rat

	t/2 (sec)		
	Acidification	Alkalinization	VC (-)
Proximal	7.43 ± 0.24 (40)	9.70 ± 0.38' (53)	7.10 ± 0.92 (9)
Distal	8.01 ± 0.54 (26)	46.2 ± 5.08' (28)	9.82 ± 1.20 (19)

' ($P<0.01$); VC (-): values obtained during voltage clamping with negative pulses.

scribed up to this point the response of the tubular epithelium to changes in transtubular pH gradients was studied. In another series of investigations, the electrochemical potential gradient for H ions was varied by an adaptation of the voltage clamp method (17). The effect of voltage changes superimposed on the normal transepithelial potential gradient can be studied by injecting current by means of an electrolyte-filled thin micropipette, and observing the effect of the ensuing voltage changes on the pH within the perfused segment. Figure 5 shows schematically the method and a graph obtained with it. Both voltage changes (represented by abrupt shifts in the recorded level) and pH modifications (given by the slower drifts in the record) are measured by the antimony microelectrode inserted into the perfusate column. Upon imposition of a potential difference which leads to a more negative lumen, a pH drift in the acid direction is observed, which reverses itself immediately when the current is turned off; from the velocity of approach to the equilibrium levels, rate coefficients equivalent to those obtained by perfusion with alkaline or acid solutions can be obtained.

Total pH changes of the order of 0.6 units per 61 mV voltage change were observed in proximal and distal tubules. The half-times of approach to the steady-state level after turning off the current are given in Table 3. It is noted that in proximal tubule the values obtained by this method are not significantly different from those found during the alkaline perfusions, which is again compatible with the pump-leak model. In the distal tubule, the difference with the alkaline perfusion method is not significant. This apparent discrepancy with the acid phosphate perfusion experiments may be due to the fact that during these voltage clamp experiments one is still working within the range of a favorable electrochemical potential gradient for H ion secretion. Furthermore, the pH variations in this type of experiment are much smaller (about 0.1 to 0.2 pH units) than those introduced during acid phosphate perfusion, where

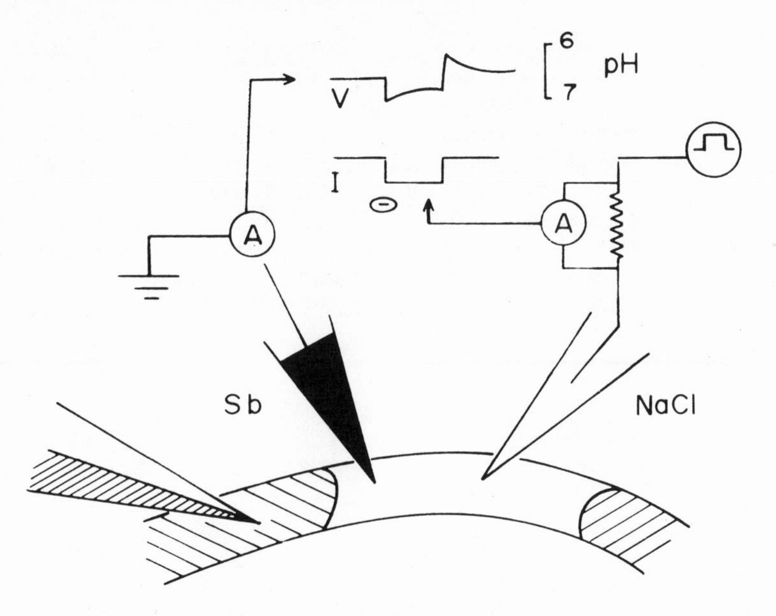

Fig. 5. Schematic drawing of "voltage clamp" experiments on renal cortical tubules. Sb, antimony microelectrode; NaCl, current passing microelectrode; A, amplifier; V, voltage recording as obtained by Sb microelectrode, representing superposition of voltage and pH changes; I, current recording.

luminal pH varies over a range of about 1 pH unit. Further studies will be necessary to clarify this point.

4. Source of H ions secreted by renal tubules.

It is commonly thought that H ions secreted into the tubular lumen originate in tubular cells, where they are either formed by dissociation of carbonic acid formed in turn by hydration of cellular CO_2, or by dissociation of water by a pump probably located at the luminal border, with subsequent neutralization of the cellular hydroxyl ions by CO_2. Several factors are thought to influence the rate of tubular acidification by acting on this mechanism, one of the more important being blood pH and pCO_2. We have used the described kinetic method in order to assess some of the factors that might be important for the tubular H ion secretory process, changing the peritubular capillaries. When these capillaries are perfused with bicarbonate Ringer's at zero CO_2 concentration, and thus at a pH of 8.4, tubular acidifying capacity is practically abolished. This experiment shows that the acidification process is closely dependent on peritubular pH and/or pCO_2, CO_2 generation within the cell being insufficient to maintain a normal rate of acidification. The importance of peritubular pH against pCO_2 was investigated in a series of experiments where peritubular capillary blood was substituted by modified Ringer's solution at different pH and pCO_2, as summarized in Figure 6. It can be seen that the steady-state buffer base level is not significantly changed with respect to controls as long as peritubular pH remains in the control range, irrespective of peritubular pCO_2; the same holds for luminal pH. When peritubular pH is altered, important changes of steady-state buffer base and pH levels occur. On the other hand, during increases in peritubular pCO_2 the acidification half-time decreases, sug-

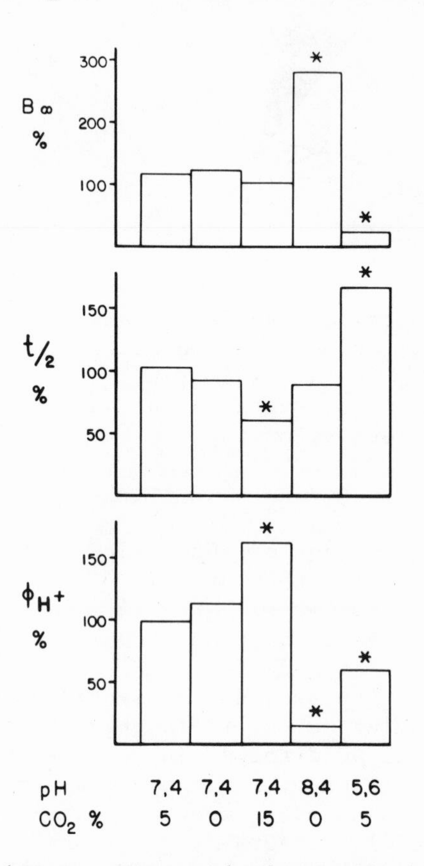

Fig. 6. Acidification parameters during peritubular perfusions with artificial solutions, in percent of control perfusions (blood in capillaries). B_∞, steady-state buffer base level; t/2 acidification half-time; ϕ_{H^+}, secretory H ion flow (or HCO_3^- reabsorption). Bottom lines: pH and CO_2 of peritubular perfusion fluid.

gesting an increase in permeability to H ions, and leading to an increase in overall acidification rate. During acid peritubular perfusions, acidification half-times increase, leading to significantly decreased secretory rates. These experiments indicate that the tubular acidifying capacity depends very markedly on peritubular pH, suggesting that capillary blood may contribute with an important fraction of the H ions transferred to the tubular lumen, or be an important regulating factor for the cellular H ion pool.

The foregoing discussion shows the potentiality of the use of antimony microelectrodes for kinetic studies on renal tubular acidification. It must be stressed, however, that the interpretation of the obtained results is still only tentative, and considerable effort will have to be made to distinguish between some of the presented alternatives, and to overcome some of the inherent draw-backs of the method. However, we believe that this method is able to bring considerable progress to the understanding of the mechanism of renal tubular acidification.

Acknowledgement: Some of the discussed aspects of renal tubular acidification have greatly benefited from discussions with Drs. T.H. Maren, G. Giebisch, F.C. Rector and B. Karlmark, during a workshop on tubular acidification held at the Department of Pharmacology and Therapeutics, University of Florida College of Medicine at Gainesville, March 1973. Work in the author's laboratory was supported by Fund. de Amparo a Pesquisa do E.S. Paulo. A.C. Cassola was a research fellow of FAPESP.

REFERENCES

1. Carrasquer, G., and Brodsky, W.A. "Elimination of Transient Secretion of Phosphate by Alkalinization of Plasma in Dogs". Am. J. Physiol. 201 (499-504), 1961.

2. Giebisch, G., Windhager, E.E., and Pitts, R.F. "Mechanism of Urinary Acidification", in Biology of Pyelonephritis. Boston, Little, Brown, 1960, p. 277-287.

3. Gottschalk, C.W., Lassiter, W.E., and Mylle, M. "Localization of urine acidification in mammalian Kidney". Am. J. Physiol. 198 (581-585), 1960.

4. Hodgkin, A.L., and Katz, B. "The Effect of Sodium Ions on the Electrical Activity of the Giant Axon of the Squid". J. Physiol. 108 (37-77), 1949.

5. Karlmark, B. "Net acid excretion from the Proximal Tubule". Acta Univ. Upsaliensis 127, 1972.

6. Karlmark, B. "The Determination of Titratable Acid and Ammonium Ions in Picomole Amounts". Anal. Biochem. 52 (69-82), 1973.

7. Levine, D.Z. "Measurement of Tubular Fluid Bicarbonate Concentration by the Cuvette-Type Glass Micro pH electrode". Yale J. Biol. Med. 45 (368-372), 1972.

8. Malnic, G., and Giebisch, G. "Mechanism of Renal Hydrogen Ion Secretion". Kidney Int. 1 (280-296), 1972.

9. Malnic, G., and Mello Aires, M. "Microperfusion Study of Anion Transfer in Proximal Tubules of Rat Kidney". Am. J. Physiol. 218 (27-32), 1970.

10. Malnic, G., and Mello Aires, M. "Kinetic Study of Bicarbonate Reabsorption in Proximal Tubule of the Rat". Am. J. Physiol. 220 (1759-1767), 1971.

11. Malnic, G., Mello Aires, M., and Giebisch, G. "Micropuncture Study of Renal Tubular Hydrogen Ion Transport in the Rat". Am. J. Physiol. 222 (147-158), 1972.

12. Malnic, G., Mello Aires, M., de Mello, G., and Giebisch, G. "Acidification of Phosphate Buffer in Cortical Tubules of Rat Kidney". Pfluegers Arch. 331 (275-278), 1972.

13. Malnic, G., and Vieira, F.L. "The Antimony Microelectrode in Kidney Micropuncture". Yale J. Biol. Med. 45 (365-367), 1972.

14. Montgomery, H., and Pierce, J.A. "The Site of Acidification of the Urine Within the Renal Tubule in Amphibia". Am. J. Physiol. 118 (144-152), 1937.

15. Rector, F.C., Carter, N.W., and Seldin, D.W. "The Mechanism of Bicarbonate Reabsorption in the Proximal and Distal Tubules of the Kidney". J. Clin. Invest. 44 (278-290), 1965.

16. Solomon, S., and Alpert, H. "A Method for Determining Titratable Acidity in Nanoliter Samples of Biological Fluids". Anal. Biochem. 32 (291-296), 1969.

17. Spring, K.R., and Paganelli, C.V. "Sodium Flux in Necturus Proximal Tubule Under Voltage Clamp". J. Gen. Physiol. 60 (181-201), 1972.

18. Steinmetz, P.R. and Lawson, L.R. "Effect of Luminal pH on Ion Permeability and Flows of Na^+ and H^+ in Turtle Bladder". Am. J. Physiol. 220 (1573-1580), 1971.

19. Uhlich, E., Baldamus, C.A., and Ullrich, K.J. "Verhalten von CO_2-- Druck und Bikarbonat im Gegenstromsystem des Nierenmarks". Pfluegers Arch. 303 (31-48), 1968.

20. Vieira, F.L. and Malnic, G. "Hydrogen Ion Secretion by Rat Renal Cortical Tubules as Studied by an Antimony Microelectrode". Am. J. Physiol. 214 (710-718), 1968.

DISCUSSION

Questions: (Brown)

1. Is the H^+ pump in tubular cells voltage-dependent?
2. Is H^+ highly permeable in these cells?

Answers: (Malnic)

1. This depends on the model you use. According to the pump-leak model, one assumes a constant pump and assumes pH changes are due to H ion backflow. However, in distal tubules the shown data may be interpreted in terms of low passive permeability and a pump dependent on the electrochemical potential gradient.

2. In the proximal tubule permeability to H ion seems to be considerable as evaluated by the rate of equilibration of alkaline and acid solutions. This would be expected from the high intercellular conductivity of this epithelium. This permeability is of the order of magnitude of that for sodium.

Question: (Fernandez)

When studying alkalinization of acidic tubular perfusates, is the rate of alkalinization changed by carbonic anhydrase inhibitors?

Answer: (Malnic)

This is an experiment we have not done yet. What I can say is that in proximal tubule alkalinization rates are faster than acidification rates after Diamox infusion, while in distal tubule these rates are approximately the same magnitude.

ELECTROCHEMICAL POTENTIALS OF POTASSIUM AND CHLORIDE IN THE PROXIMAL RENAL TUBULES OF NECTURUS MACULOSUS

Raja N. Khuri

Department of Physiology

American University of Beirut, Beirut, Lebanon

INTRODUCTION

The renal tubular epithelium is essentially a 3-compartment system: interstitial, intracellular and luminal. These three aqueous phases are separated by two lipid plasma membranes: the outer peritubular cell membrane and the inner luminal cell membrane.

Knowledge of the mechanism of ion transport across a segment of the renal tubular epithelium like the proximal convoluted tubule is based on our ability to quantitate both the electrical and chemical driving forces for a given ionic species operating across the different boundaries. Electrical gradients across cell membranes can be measured with reasonable confidence. However, not before first resolving the uncertainties about intracellular ionic activities or concentrations can we quantitate chemical gradients.

The aim of this study is to define the electrochemical driving forces for the movement of potassium and chloride in renal proximal tubules. The approach involved the simultaneous and direct potentiometric determination of electrical PDs and the intracellular potassium and chloride concentration of single cellular elements of the proximal tubular epithelium. Indirect methods cannot resolve the uncertainties concerning intracellular ions, their possible binding and compartmentalization.

Potassium is the major determinant of the electrophysiological properties of cells- renal epithelial cells being no exception (Windhager and Giebisch, 1965). This study utilizes a double-barreled K^+-selective liquid ion-exchange microelectrodes to measure effective intracellular K^+ concentration in single cellular elements of the proximal tubular epithelium.

This study was performed on the Necturus kidney proximal tubule because of the number of advantages that this preparation offers. The relatively large size of the tubular cells renders electrometric analysis simpler. The basal infoldings of the proximal tubular epithelium of Necturus do not obliterate the intracellular space as they do in this segment of the mammalian nephron. The absence of respiratory and pulsatile movements permits more stable cellular impalements.

The value of the mean intracellular K^+ activity was significantly lower than the total K^+ content. However, there was remarkable agreement in the magnitudes of the calculated K^+ equilibrium potentials across the luminal and peritubular cell boundaries on the one hand and their respective measured membrane PDs on the other hand. This implies, that as in skeletal muscle, K^+ is in electrochemical equilibrium distribution across the boundaries that separate the different compartments of the proximal tubular system.

Sodium and chloride are the major osmotic constituents of glomerular filtrate that need to be conserved by reabsorptive transport across the renal tubular epithelium. Double-barreled Cl^--selective liquid ion-exchange microelectrodes were used to measure intracellular Cl^- concentration in single cellular elements of the proximal tubular epithelium.

The electrometric intracellular $[Cl^-]$ of 18.7 ± 1.3 mM, while it accounts for only 2/3 of the total Cl^- content of proximal tubule cells, is still significantly greater than that expected from a simple passive distribution of this ion between the intracellular fluid and the two extracellular fluid compartments (luminal and peritubular). Therefore, chloride must be actively transported across the luminal membrane by an anionic pump or a neutral NaCl pump. This constitutes the first or luminal step in transcellular chloride reabsorptive transport. In the second or, peritubular step, Cl^- could passively accompany the actively and electrogenically extruded Na^+ as well as be a component of a peritubular electroneutral NaCl active transport process.

The renal tubular epithelium consists of three aqueous compartments separated by two lipid membranes (luminal and peritubular). Across these two boundaries ionic potassium, but not chloride, is distributed in accordance with the Donnan relationship of electrochemical equilibrium.

METHODS

Electrometric Methods

The process of constructing ion-selective double-barreled liquid ion-exchange microelectrodes follows the general guidelines outlined for the construction of the single-barreled micropipette (Khuri *et al.*, 1971; Khuri *et al.*, 1972). Two segments of Pyrex capillary tubing, cemented together, are held over a microflame and at the softening point rotated 360° and pulled slightly. Then it is mounted in a vertical pipette puller and pulled into two double-barreled micropipettes with tip diameter of one micron or less. Fig. 1 is a diagrammatic representation of the electrode.

The K^+-selective liquid ion exchanger (Corning Catalog No. 476132) and the Cl^--selective liquid ion exchanger are organic electrolytes dissolved in organic solvent. The exchangers are substantially insoluble in aqueous solutions. Their adhesiveness to non-siliconized glass surfaces is poor in comparison with water which can readily displace them. Therefore, the tip end of the shank of the exchanger (indicator) barrel is rendered hydrophobic-organophilic by siliconization with 1.25 % silicone oil (Dow Corning) in trichlorethylene. The tip of the reference barrel is not siliconized. Then the micropipettes are cured in an oven at 200° C for two hours.

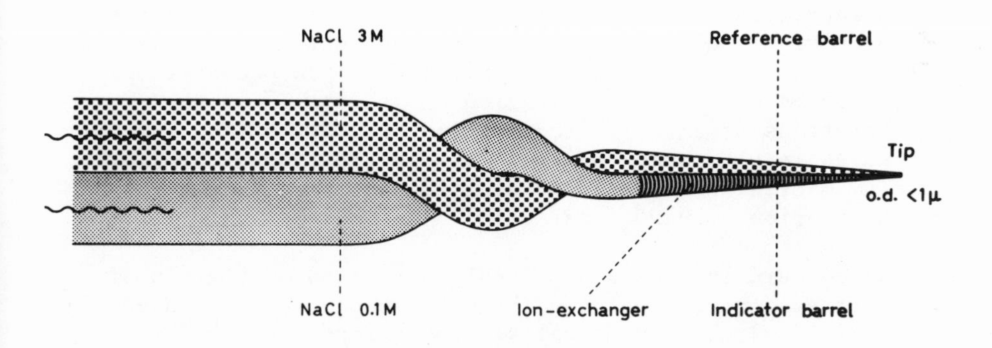

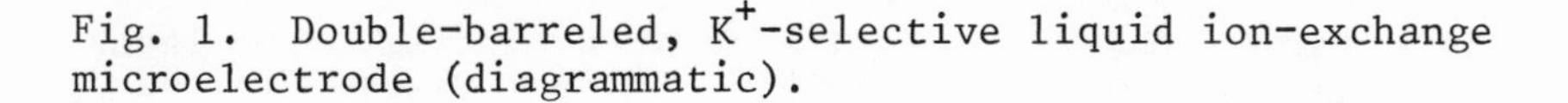

Fig. 1. Double-barreled, K^+-selective liquid ion-exchange microelectrode (diagrammatic).

After siliconization the two barrels are filled separately with the ion-exchanger and the reference salt solution by means of filler microcapillaries. The rest of the shank and the step of the exchanger barrel is filled with 0.1 M NaCl. An Ag-AgCl wire is inserted into the stem of each barrel as an internal reference element.

The potential measurements were made by means of a pair of electrometers and the readings were displayed on a Grass polygraph. All equipment was placed inside a Faraday cage and grounded to a common copper bar. Potassium microelectrodes were considered suitable for use if they met the following criteria. A high potassium sensitivity exhibited by a response of 55 to 60 mV/decade change of ionic activity, a K^+ : Na^+ selectivity coefficient of better than 40:1, a low tip potential of < 5 mV and minimal drift of potential with time of < 1 mV/h. A representative potassium microelectrode had a lag time of 18 msec, a time constant of 85 msec and a rise time (from 10 % - 90 % of final value) of 160 msec. Chloride microelectrodes were considered suitable for use if they met a number of criteria. An adequate chloride sensitivity exhibited by a response of 50 or more mV/decade change of chloride concentration, selectivity coefficients of Cl^- : HCO_3^- of 25:1 and Cl^- : $H_2PO_4^-$ of greater than 10:1, a low tip potential of < 5 mV and a minimal drift of potential with time of < 1 mV/hour. A representative chloride microelectrode had a lag time of 20 msec, a time constant of 100 msec and a rise time of 200 msec.

The selectivity coefficients of the microelectrodes were evaluated in equimolar pure solutions. The ionic strength of the standards was selected so as to approximate the value of intracellular fluid of Necturus proximal tubules. The intracellular K^+ and Cl^- concentrations were determined from the calibration curve of the respective microelectrode in standard solutions that flanked intracellular readings. The chloride microelectrode potential is insensitive to the cationic substitution of Na^+ for K^+ in the different solutions at constant ionic strength. It is also insensitive to changes in H^+ concentration in the pH 4-8 range. Similarly, the potassium microelectrode potential is insensitive to the anionic substitution of $H_2PO_4^-$ for Cl^-.

A major concern in this study was localization of the tip of the double-barreled microelectrode (Fig. 1) in the intracellular compartment of proximal tubular cells. The reference barrel is the electrical sensor; the other barrel being the chemical (potassium or chloride) sensor. The leads of the two barrels were connected to one electrometer and this gave the ionic potential. The leads of the reference barrel and an external 3 M NaCl single pipette were connected to another electrometer and the PD between them represented the membrane potential. The ionic potential (K^+ or Cl^-) and the membrane potential were measured and recorded simultaneously on a Grass polygraph.

The double-barreled electrode was mounted on an Electrode Carrier connected to a Hydraulic Micro-Drive (David Kopf, model 1207 S). The electrode Carrier itself is mounted on a Leitz micromanipulator. Initially, the tips of the double-barreled microelectrode and the single-barreled micropipette are immersed in Ringer solution covering the kidney. The tip of the double-barreled microelectrode is advanced very close to the peritubular surface of an identified latter proximal tubular segment. Subsequent advances are made in a stepwise fashion few micra at a time by means of the remote control mechanism. As the double-barreled microelectrode impales the peritubular membrane of a Necturus proximal tubule (Fig. 2a), there is an abrupt steep rise in the membrane potential (upper tracing) of 70-80 mV associated with a more gradual rise in the potassium potential. After recording a stable peritubular membrane PD and a cell K^+ potential, the double-barreled electrode

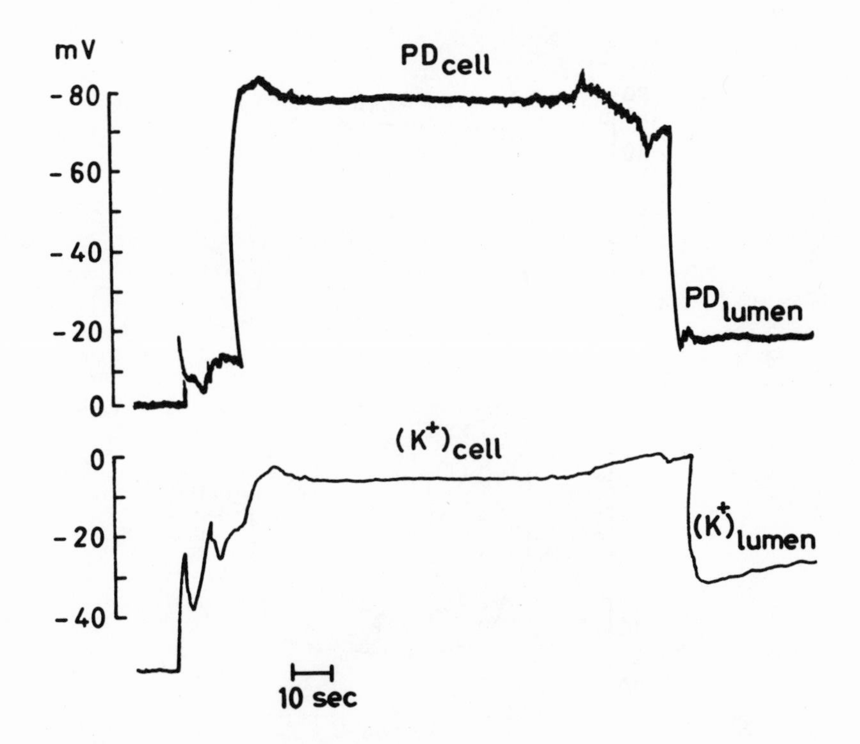

Fig. 2a. Simultaneous recording of membrane PDs (upper) and potassium potentials (lower) by the two barrels of the double-barreled K^+-selective microelectrode.

was advanced a few micra into the lumen of a late proximal tubule to record two stable potentials: a transepithelial PD of about 12-15 mV and a tubular fluid K^+ potential equivalent to about twice the plasma K^+ activity.

Fig. 2b represents a simultaneous recording obtained with a Cl^--selective microelectrode. As the double-barreled microelectrode impales the peritubular membrane, there is a steep rise in the membrane potential of about - 70 mV associated with a fall in the Cl^- potential. The fall in the chloride potential is less steep than the rise in the peritubular membrane PD. This is due to the fact that the chemical sensor has a slower response time than the electrical cell. After recording a stable peritubular membrane PD and cell Cl^- potential, the double-barreled electrode was advanced a few micra into the lumen of a late proximal tubule to record two stable potentials: a transepithelial PD of about 10-15 mV and a tubular fluid Cl^- potential roughly equivalent to the Cl^- potential in the amphibian Ringer's solution covering the kidney.

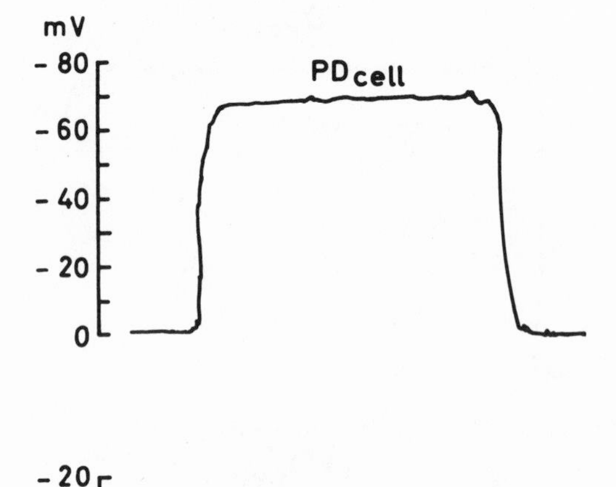

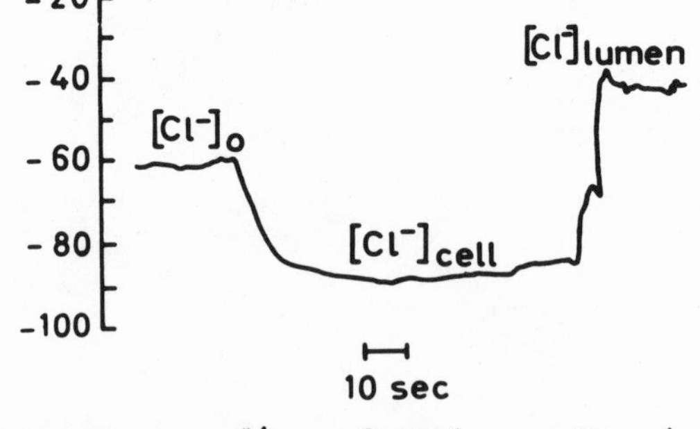

Fig. 2b. Simultaneous recording of membrane PDs (upper) and chloride potentials (lower) by the two barrels of the double-barreled Cl^--selective microelectrode.

Renal Methods

Adult Necturus maculosus, flown to Beirut between December and March, were used within a month of their arrival. The Necturii were maintained unfed in tap water in a cold room at 4° C. The Necturii were anesthetized by immersion in tap water containing 0.1 % tricaine methanesulfonate. The gills were submerged in water into which air was continuously bubbled. The left kidney was exposed. Amphibian Ringer in 2 % agar was poured over the kidney. A hole in the center of the agar was filled with aqueous amphibian Ringer solution having a $[K^+]$ of 3.6 mM and a $[Cl^-]$ of 75 mM. The extracellular reference micropipette was immersed in this solution. All the experiments were carried out at room temperature (about 20° C).

The latter part of the proximal tubule of Necturus is straight and superficial. It tapers to form the thin intermediate segment as it courses medially towards the glomerulii and central vein. This segment could be identified with lisamine green infusion or Sudan black colored mineral oil luminal injections. Free flow collections for fluid/plasma ratio of inulin-^{14}C and the direct potentiometric determinations of K^+ in intracellular and luminal fluid were made in the latter part of the proximal tubule.

The thick, club-shaped caudal mass of kidney was cut into two slices. The lateral and medial slices are placed on tin foil and the wet weight quickly obtained using a Mettler balance. The dry weight was determined after drying in the oven for 2 h at 110° C. The dried pieces were digested in concentrated nitric acid. The K^+ concentration of the dissolved tissue was determined by flame photometry. The chloride concentration was determined by potentiometric titration (Ramsay *et al.*, 1955). Aliquots of the slices were taken in the fresh state and incubated at room temperature in amphibian Ringer solution containing inulin-^{14}C for 45 minutes in a Dubnoff Metabolic shaking incubator for extracellular space determination.

RESULTS AND DISCUSSION

Potassium

To check the performance and accuracy of the potassium microelectrode serial readings were made in the K^+ calibrating standards. This yielded a mean ratio of potentiometric / photometric K^+ concentration of 1.01 ± 0.02 (S.E.), a ratio which is not significantly different from unity. In addition the performance of the potassium microelectrode was tested in Necturus and rat serum. The K^+ concentration was measured in Necturus and rat serum droplets *in vitro* and the values obtained were compared with the flame photometric K^+ concentration as determined on an aliquot sample. This yielded a mean ratio of potentiometric / photometric K^+ concentration of 1.00 ± 0.02 (S.E.), a value which is not significantly different from unity.

Table I gives a summary of the water and cation composition of lateral slices of the Necturus kidney. The mean values for the K^+ and Na^+ concentrations of medial slices were not significantly different from the values of lateral slices given in the table. This is in agreement with Whittembury *et al.* (1961). In order to test the postulate that the smaller distal tubules of medial slices can pack better than the larger proximal tubules of lateral slices, thereby leaving a smaller interstitial space, extracellular space determinations by inulin-^{14}C were carried out on pairs of lateral-medial slices of the same kidneys. The mean extracellular space of 20.0 ± 0.6 % of lateral slices is not significantly different from the mean of 18.0 ± 0.9 % of medial slices. Thus disproving our postulate that the interstitial space is a function of tubular packing.

This electrometric study yielded a mean intracellular K^+ activity of 58.7 ± 2.3 mM. Defining an experimental mean ionic activity coefficient as the ratio of electrometric K^+ activity over photometric K^+ concentration (103 mM per kg cell water), we obtain for our data a value of 0.57. This value is much lower than the mean cationic activity coefficient of 0.77 predicted for the ionic strength of Necturus body fluids. If from this one were to conclude that about 1/4 of the total intracellular K^+ content is bound, this

TABLE I

Water and Ionic Composition of Lateral Slices (Proximal Tubules) of Necturus Kidney

Water	ECF	K^+ (mM)	Na^+ (mM)	Cl^- (mM)
Percent of Wet Weight		Per kg Cell Water		
84.0	20.0	103.0	38.1	32.1
±0.2	±0.6	±1.8	±1.8	±2.3

would contrast with nerve (Hinke, 1961) and skeletal muscle (Lev, 1964) where measurements with K^+ glass microelectrodes revealed that all the intracellular K^+ is in free solution. Our electrometric evidence for binding and/or compartmentalization of the proximal tubule cell K^+ is consistent with the previously reported (Mudge, 1953; Wiederholt *et al.*, 1971) radio-kinetic observations that a significant fraction of kidney cortex K^+ is not readily exchangeable and that the intracellular K^+ pool is not homogeneous.

An intracellular K^+ activity of 58.7 is equivalent to a concentration of about 76 mM. If between 1/3 to 1/2 of the total intracellular Na^+ is free, then the intracellular fluid of the proximal tubular epithelium would have an osmolarity and an ionic strength quite similar to both the peritubular and luminal fluid bathing its two boundaries.

Table II gives a summary of the late proximal tubule activity ratios, electrical PDs and calculated K^+ equilibrium potentials. These parameters are presented under three headings: (1) across the tubular epithelium (transepithelial), (2) across the luminal cell membrane (luminal), and (3) across the peritubular cell membrane (peritubular). Whereas the transepithelial treatment is essentially one that deals with a two-compartment system, analysis across the luminal and peritubular boundaries involves a three-compartment system.

The first section of Table II represents the transepithelial parameters. The fluid/plasma K^+ activity ratio of 1.9 ± 0.2 is significantly greater than unity ($P < 0.01$) but is not significantly different from an inulin ratio of 1.5 ± 0.2 in the latter part of the proximal tubule of the Necturus kidney. The potassium equilibrium potential (E_K) calculated from the Nernst equation using the K^+ activity ratio yield a value of 16.4 ± 1.0 mV. This is not significantly different from the mean transepithelial PD of 14.5 ± 1.5 mV. Thus K^+ ion is in electro-chemical equilibrium distribution across the proximal tubular epithelium of Necturus. This is consistent with the fact that in Necturus the proximal tubule does not modify the filtered potassium load. There is no significant net K^+ transport, reabsorptive or secretory, in the Necturus proximal tubule. The observed electrical asymmetry of this epithelium may be due to the difference in magnitude of two K^+ diffusion potentials in series. Thus the simultaneously observed K^+ activity distribution ratios can account for all the electrical asymmetry without the need to invoke a significant effect of passive leak of Na^+ on either membrane boundary.

At the luminal cell boundary (Table II, Fig. 3) the measured membrane potential (E_m) of 57.5 ± 1.5 mV and the calculated K^+ equilibrium potential (E_K) of 58.9 ± 1.0 mV are not significantly

TABLE II

Late Necturus Proximal Tubule Activity Ratios, Electrical Potential Differences, and Calculated K^+ Equilibrium Potentials Across the Tubular Epithelium (Transepithelial) and the Inner (Luminal) and Outer (Peritubular) Cell Boundaries. Mean Values ± Standard Error are Given. All Values Given are Derived from Purely Electrometric Determinations.

Transepithelial					Luminal					Peritubular				
$(TF)_K$ mM	$(P)_K$ mM	$(TF/P)_K$	E_K mV	E_m mV	$(Cell)_K$ mM	$(TF)_K$ mM	$(Cell/TF)_K$	E_K mV	E_m mV	$(Cell)_K$ mM	$(P)_K$ mM	$(Cell/P)_K$	E_K mV	E_m mV
5.4	2.8	1.9	16.4	14.5	58.7	5.4	10.9	58.9	57.5	58.7	2.8	21.0	75.4	72.0
±0.1	±0.3	±0.2	±1.0	±1.5	±2.3	±0.1	±0.4	±1.0	±1.5	±2.3	±0.3	±0.8	±1.0	±1.4

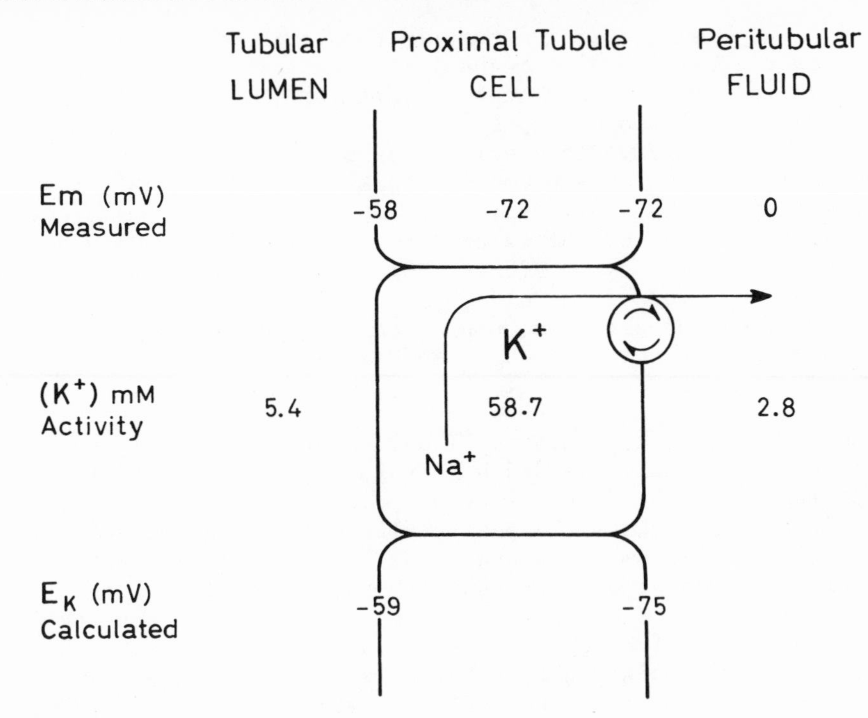

Fig. 3. A schematic representation of K^+ and Na^+ transport in the three-compartment system of the proximal tubule of Necturus kidney.

different. Thus potassium ion is in an electro-chemical equilibrium distribution across the luminal cell border. In contrast to K^+ which has no net movement, Na^+ is driven across the luminal membrane passively by an electro-chemical gradient.

At the peritubular cell boundary (Table II, Fig. 3) the measured membrane potential (E_m) of 72.0 ± 1.4 mV and the calculated K^+ equilibrium potential (E_K) of 75.4 ± 1.0 mV are not significantly different. Thus potassium ion is also in an electro-chemical equilibrium distribution across the peritubular cell border. In the absence of any net movement of K^+ across the peritubular membrane we do not need to postulate the presence of an active K^+ influx component as part of the peritubular Na^+ pump.

Fig. 3 shows a single sodium pump in the peritubular cell membrane since the Na extrusion mechanism is not necessarily linked with K^+ uptake into the cell. The intracellular $[K^+]$ can all be accounted for on the basis of a passive electro-chemical equilibrium distribution of potassium. The peritubular membrane is known (Giebisch, 1961) to be highly permeable to K^+ ion and to exhibit a high K^+ selectivity. The observation that the peritubular membrane of the Necturus proximal tubule is characterized by $E_K = E_m$ may be explained in one of two ways. Either that the PD across the peritubular cell membrane is mainly due to a K^+ diffusion potential, or that the peritubular cationic pump is an electrogenic Na^+ pump which by its operation generates E_m. Then a passive K^+ influx occurs to the point where $E_K = E_m$.

In conclusion, intracellular $[K^+]$ as measured indirectly by chemical methods generally yielded high values which resulted in the overestimation of the K^+ equilibrium potential. To account for an E_K greater than E_m across the peritubular cell boundary, an active K^+ influx mechanism was postulated. However, the direct electrometric measurement of intracellular potassium yields the true value of E_K. This study has shown that $E_K = E_m$ for both the peritubular and the luminal cell membranes. In other words, both the luminal and peritubular cell membranes act as potassium electrodes. Therefore, as in skeletal muscle (Khuri *et al.*, 1972), a Donnan-type equilibrium holds with respect to potassium ion passive distribution across both boundaries.

Chloride

Serial readings were made in the calibrating standards to check the performance and accuracy of the chloride microelectrode. This yielded a mean ratio of potentiometric / analytical Cl^- concentration of 1.02 ± 0.02 (S.E.), a ratio which is not significantly different from unity. The performance of the microelectrode was checked in serum. The direct potentiometric values obtained were compared with the Cl^- concentration as determined by a colorimetric method. This yielded a mean ratio of potentiometric / colorimetric Cl^- concentration of 1.03 ± 0.02 (S.E.), a value which is not significantly different from unity.

Table 1 gives the mean value for the chloride total concentration of lateral slices of the Necturus kidney as determined by potentiometric titration. The mean chloride concentration of 32.1 ± 2.3 mM per Kg cell water is in very close agreement with Whittembury *et al.* (1961).

This electrometric study yielded a mean intracellular Cl^- concentration of 18.7 ± 1.3 mM per Kg of proximal tubule cell water.

This accounts for about 2/3 of the chloride content of proximal tubule cells. The discrepancy between the active electrometric $[Cl^-]$ of 18.7 mM and the total intracellular chloride of 32.1 mM may be accounted for by one of three possibilities. First, a non-uniform distribution of chloride within different 'intracellular' compartments. The lateral intercellular space which may not be readily penetrated by inulin is a region of hypertonicity rich in chloride. Second, 1/3 of the intracellular chloride may be bound to proteins or other macromolecules or fibrous elements. Third, cells of distal tubules may have a higher intracellular chloride (Conway *et al.*, 1946) than those of proximal tubules.

Still the electrometrically determined intracellular $[Cl^-]$ of 18.7 mM is too high and too far from an equilibrium distribution to be consistent with passive transport in the three-compartment system. Thus intracellular chloride is at a higher electrochemical potential than either luminal or peritubular fluid chloride. This results in the passive efflux of intracellular Cl^- across the luminal and peritubular cell membranes. Since the tubule cell has greater $[Cl^-]$ than that expected from a simple passive distribution (5-7 mM) between intracellular and the two extracellular fluid compartments, Cl^- must be actively transported into the cell.

Table III gives a summary of the late proximal tubule concentration ratios, electrical PDs and calculated Cl^- equilibrium potentials. These parameters are presented under two headings: (1) across the luminal membrane (luminal), and (2) across the peritubular cell membrane (peritubular). The analysis across the luminal and peritubular boundaries involves a three-compartment system.

Table IV summarizes the algebraic summation of electrical and chemical driving forces yielding the net electrochemical force (i.e. potential gradient in mV) which drives chloride ions across the luminal and peritubular cell boundaries. Fig. 4 is a schematic representation of chloride transport in the three-compartment system of the proximal tubule of the Necturus kidney.

At the luminal cell boundary (Table III, Fig. 4) the measured membrane potential (E_m) of 58.5 ± 1.5 mV is significantly ($P < 0.001$) greater than the calculated Cl^- equilibrium potential (E_{Cl}) of 38.0 ± 2.2 mV. For an anion an $E_m > E_{Cl^-}$ implies active Cl^- influx. Since the electrochemical PD which drives chloride ions from the cell into the lumen is about 20.5 mV (Table IV), the luminal membrane active chloride reabsorptive mechanism must exceed 20 mV in order to reabsorb Cl^- out of the lumen. The net electrochemical force which moves Cl^- from cell-to-lumen accounts for the observed Cl^- flux in the stop-flow experiments of Kashgarian *et al.* (1965).

TABLE III

Late Necturus Proximal Tubule Concentration Ratios, Electrical Potential Differences, Calculated Cl^- Equilibrium Potentials Across the Inner (Luminal) and Outer (Peritubular) Cell Boundaries. Mean Values ± Standard Error are Given. All Values Given are Derived from Purely Electrometric Determinations.

Luminal					Peritubular				
$[Cell]_{Cl}$ mM	$[TF]_{Cl}$ mM	(Cell/TF)	E_{Cl} mV	E_m mV	$[Cell]_{Cl}$ mM	$[P]_{Cl}$ mM	$(Cell/P)_{Cl}$	E_{Cl} mV	E_m mV
18.7	82.5	0.22	38.0	58.5	18.7	78.9	0.24	43.8	70.2
±1.3	±3.6	±0.02	±2.2	±1.5	±1.3	±3.6	±0.02	±2.5	±1.3

TABLE IV

Electrical and Chemical Driving Forces (Potential Gradients) Across the Luminal and Peritubular Boundaries and Their Algebraic Summation.

	Luminal Membrane mV	Peritubular Membrane mV
Electrical driving force	58.5	70.2
Chemical driving force	38.0	43.8
Net electro-chemical force	20.5	26.4
	favoring Cl^- efflux: cell-to-lumen	favoring Cl^- efflux: cell-to-peritubular

The luminal membrane must possess an active mechanism capable of transporting Cl^- against an electrochemical potential gradient. This active chloride reabsorptive transport in the luminal membrane could be either a Cl^- ionic pump or a neutral NaCl pump. Na^+ may be driven across the luminal membrane passively along a combined chemical and electrical potential gradients. This study cannot differentiate between a lumen-to-cell passive Na^+ influx electrically coupled to an active Cl^- influx and an electrically neutral luminal NaCl pump involving a double carrier that can combine with Na and Cl simultaneously as proposed by Diamond for the gall bladder (1962). A third possibility is that part of the free energy dissipated by the passive influx of Na^+ from lumen-to-cell down a combined electro-chemical gradient, is utilized to energize the uphill transport of Cl^- from lumen-to-cell.

At the peritubular cell boundary (Table III, Fig. 4) the measured membrane potential (E_m) of 70.2 ± 1.3 mV is significantly ($P < 0.001$) greater than the calculated Cl^- equilibrium potential of 43.8 ± 2.5 mV. Thus there is a net electro-chemical PD or force of 26.4 mV (Table IV) favoring Cl^- efflux from cell-to-peritubular interstitium. In contrast to chloride, Na^+ movement from cell-to-interstitium across the peritubular cell membrane must proceed uphill against both a chemical and an electrical potential gradient. This active Na^+ transport may be effected by two mechanisms. First, an electrogenic Na^+ pump whose active extrusion of Na^+ is accompanied

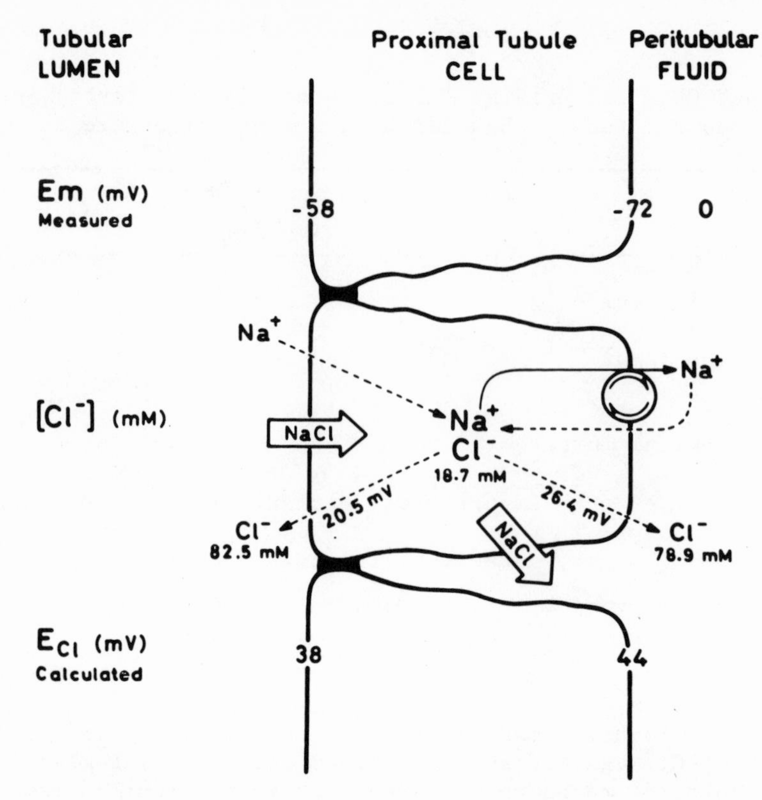

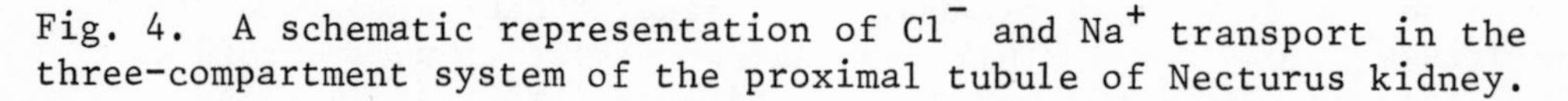

Fig. 4. A schematic representation of Cl^- and Na^+ transport in the three-compartment system of the proximal tubule of Necturus kidney.

by the passive extrusion of Cl^-. Such a homocellular (Maude, 1970) transport mechanism would regulate cell sodium and ionic composition. Second, an electroneutral NaCl pump which deposits salt either into the hypertonic lateral intercellular space or directly on the basal side of the epithelium. Such a transcellular (Maude, 1970) transport mechanism would contribute to net salt reabsorption across the tubular epithelium. Sodium influx from the peritubular fluid into the cell is a passive process (Fig. 4) driven by a sizable electrochemical potential gradient of about 100 mV.

Potassium Chloride

Conway *et al.* (1946) concluded that the cells of the frog proximal tubule, like frog sartorius muscle, are freely permeable to K and Cl and that these two ions are distributed in accordance with the Donnan relationship of electrochemical equilibrium. This means that the intracellular and extracellular products of K and Cl concentrations are equal.

That is, $[K]_o \cdot [Cl]_o = [K]_i \cdot [Cl]_i$

But our values, $3.6 \times 78.9 < 76 \times 18.7$

And, therefore $[K]_o \cdot [Cl]_o = 0.2[K]_i \cdot [Cl]_i$

Thus a coefficient of 0.2 equalizes the products of tissue and medium K^+ and Cl^- concentrations. It is the chloride ratio Cl_o/Cl_i or distribution and not the potassium distribution which is responsible for this departure from the Donnan electrochemical equilibrium distribution.

SUMMARY

Using double-barreled liquid ion-exchange microelectrodes intracellular potassium and chloride concentrations were measured simultaneously with membrane PDs in single cells of the proximal tubules of Necturus kidney. The electrometric method measured an intracellular potassium concentration which constituted about 3/4 of the total K^+ content. There was a remarkable agreement between the calculated E_K and measured E_m across the luminal and peritubular boundaries suggesting passive transport of K^+, an electrochemical equilibrium distribution and K^+-selectivity of both membranes. The electrometric intracellular chloride measured 2/3 of the total Cl^- content. Chloride was not distributed in accordance with the Donnan relationship since $E_m > E_{Cl}$ across both membranes. The latter is evidence for active transport of Cl^- across the luminal membrane either by means of an ionic Cl^- pump or a neutral NaCl pump.

REFERENCES

CONWAY, E. J., FITZGERALD, D., & MACDOUGALD, T. C. Potassium accumulation in the proximal tubules of the frog's kidney. *Journal of General Physiology*, 1946, *29*, 305-334.

DIAMOND, J. M. The mechanism of solute transport by the gall-bladder. *Journal of Physiology* (London), 1962, *161*, 474-502.

GIEBISCH, G. Measurement of electrical potential differences on single nephrons of the perfused kidney. *Journal of General Physiology*, 1961, *44*, 659-678.

HINKE, J. A. M. The measurement of sodium and potassium activities in the squid axon by means of cations-selective glass microelectrodes. *Journal of Physiology* (London), 1961, *156*, 314-355.

KASHGARIAN, M. H., WARREN, Y., & LEVITIN, H. Micropuncture studies of renal tubular excretion of chloride during hypercapnia. *American Journal of Physiology*, 1965, *209*, 655-659.

KHURI, R. N., AGULIAN, S. K., & WISE, W. M. Potassium in the rat kidney proximal tubules *in situ*: determination by K^+-selective liquid ion-exchange microelectrodes. *European Journal of Physiology*, 1971, *322*, 39-46.

KHURI, R. N., HAJJAR, J. J., & AGULIAN, S. K. Measurement of intracellular potassium with liquid ion-exchange microelectrodes. *Journal of Applied Physiology*, 1972, *32*, 419-422.

LEV, A. A. Determination of activity coefficients of potassium and sodium in frog muscle fibers. *Nature* (London), 1964, *201*, 1132-1134.

MAUDE, D. C. Mechanism of salt transport and some permeability properties of rat proximal tubule. *American Journal of Physiology*, 1970, *218*, 1590-1595.

MUDGE, G. H. Electrolyte metabolism of rabbit kidney slices: studies with radioactive potassium and sodium. *American Journal of Physiology*, 1953, *173*, 511-522.

RAMSAY, J. A., BROWN, R. H. J., & CROGHAN, P. C. Electrometric titration of chloride in small volumes. *Journal of Experimental Biology*, 1955, *32*, 822-829.

WHITTEMBURY, G., SUGINO, N., & SOLOMON, A. K. Ionic permeability and electrical potential differences in Necturus cells. *Journal of General Physiology*, 1961, *44*, 689-712.

WIEDERHOLT, M., SULLIVAN, W. J., GIEBISCH, G., SOLOMON, A. K., & CURRAN, P. F. Measurement of unidirectional K and Na fluxes on single distal tubules of Amphiuma kidney. *Journal of General Physiology*, 1971, *57*, 495-525.

WINDHAGER, E. E., & GIEBISCH, G. Electrophysiology of the nephron. *Physiological Reviews*, 1965, *45*, 214-244.

IV. Brain

SOME MEASUREMENTS OF EXTRACELLULAR POTASSIUM ACTIVITY IN THE MAMMALIAN CENTRAL NERVOUS SYSTEM

Mary E. Morris and K. Krnjević

Department of Research in Anaesthesia

McGill University, Montreal 101, Canada

Although changes in extracellular K^+ activity (a_K) have long been suspected in mechanisms of function and dysfunction in the central nervous system (Dubner & Gerard, 1939; Grafstein, 1956; Kuffler, 1967), only recently has it become practicable to record them both directly and with precise localization (Hinke, 1961; Walker, 1971). We have used micro-electrodes containing a liquid ion-exchanger (Walker, 1971) to measure extracellular a_K in the cerebral cortex, and also in regions of the spinal cord and the medulla which receive the terminals of somato-sensory nerve fibres (Krnjević & Morris, 1972).

METHODS

Our K-sensitive electrodes were prepared by the method described by Walker (1971), except that the tips were coated over a distance of 400-1000 μm with 3,3,3-(trifluoropropyl)trimethoxysilane (Pierce Chemical Co.). The K-liquid ion exchanger (Corning Code 477317) was introduced into the tip from above by injection from a syringe and the application of pressure. The electrodes had tip diameters of 1-1.5 μm and resistances of 10^9-10^{10} Ω.

Reference micro-electrodes were filled with 1 M NaCl or 1 M NaCl in 1% (w/v) agar (Fisher Scientific Co.), and had tips of 1-10 μm diameter and resistances of 2-10 MΩ. In order to distinguish K^+ potential changes (ΔE_K) in the presence of other tissue potential changes (ΔV) generated by cellular activity, the reference micro-electrode was placed very close to the K-sensitive electrode. This gave reasonable rejection (>90%) of common mode signals (see Fig. 4).

The reference micro-electrode was bent near the tip before filling, so that the distal part of the shank could be aligned approximately parallel to that of the ion-exchanger electrode (see Fig. 1). This increased the mechanical stability of the double electrode and caused less damage during penetration of the tissue. The electrodes were held in close approximation by mounting them in a Narishige electrode holder and adjusting the tips in three planes under microscopic control until they were separated by a gap of 5-50 μm (Krnjević & Schwartz, 1967); they were then fixed permanently with epoxy glue and dental cement, and stored at 4°C with the tips in distilled water.

The potentials of the ion-exchanger and reference electrodes were led via guarded cables to a model 604 Keithley electrometer (input impedance > 10^{14} Ω) (see Fig. 2). The K potential (E_K) was recorded differentially, and the voltage of the reference electrode with respect to ground amplified separately. Both signals (E_K and V) were recorded on a Grass polygraph. In some experiments E_K was electronically converted to its antilog to provide an output more linearly related to a_K (see Fig. 2).

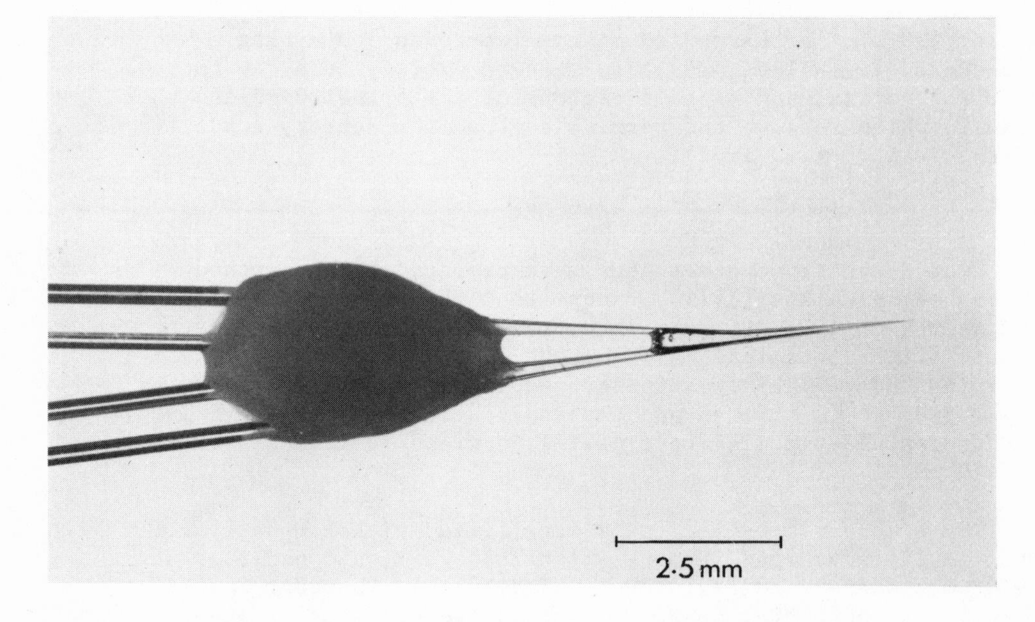

Fig. 1. Photograph of twin potassium electrode assembly. Reference electrode angled near tip, positioned close to potassium selective electrode and attached with epoxy and dental cement.

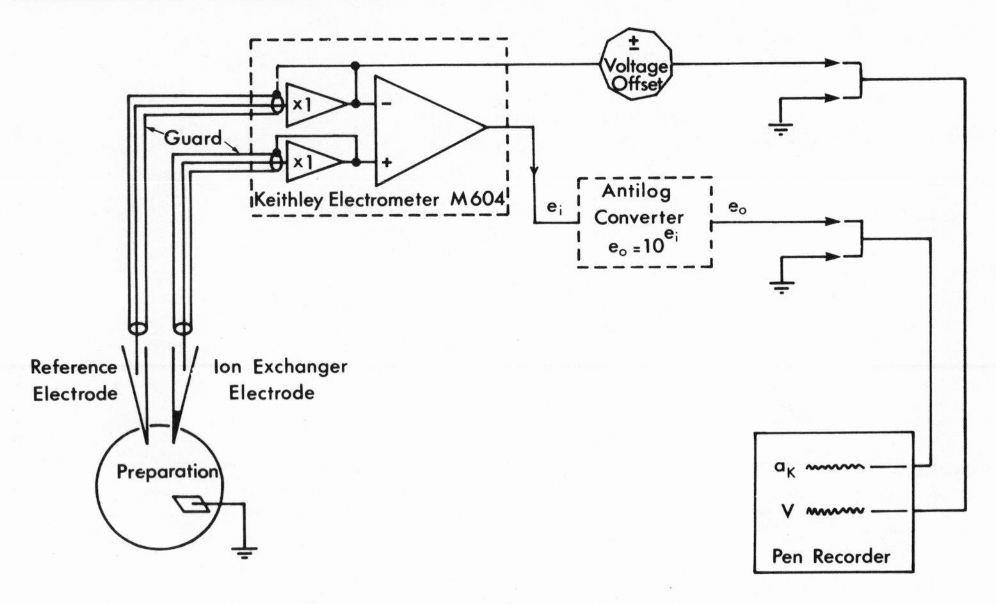

Fig. 2. Schematic diagram of recording arrangement. Voltage signals from K and reference micro-electrodes amplified with a differential electrometer. Negative input from reference electrode records tissue potential changes (V) during neuronal activity. Potassium potential recorded differentially (E_K) or electronically converted to its antilog, approximating a_K.

When calibrations were made in solutions whose composition (NaCl 137 mM, $MgCl_2$ 1.3 mM, $CaCl_2$ 1.5 mM, NaH_2PO_4 1 mM, $NaHCO_3$ 11.9 mM) resembled cerebrospinal fluid (Ames, Higashi & Nesbett, 1965), the slope of the electrode response to changes in [K^+] was greater at higher concentrations of KCl; in the example shown in Fig. 3, there was a response of 44 mV between 2 and 20 mM K, and of 55 mV between 10 and 100 mM K.

The full response time of the electrode to a 100-fold change in potassium concentration, when estimated roughly by recording E_K levels before and after the immersion of a droplet of high concentration in a large volume of low concentration, was < 0.3 sec.

Control tests were made to compare the voltage recorded differentially by the K and reference electrodes, and that by the reference electrode alone. Fig. 4A shows potentials recorded in the dorsal horn in one animal experiment in response to repetitive

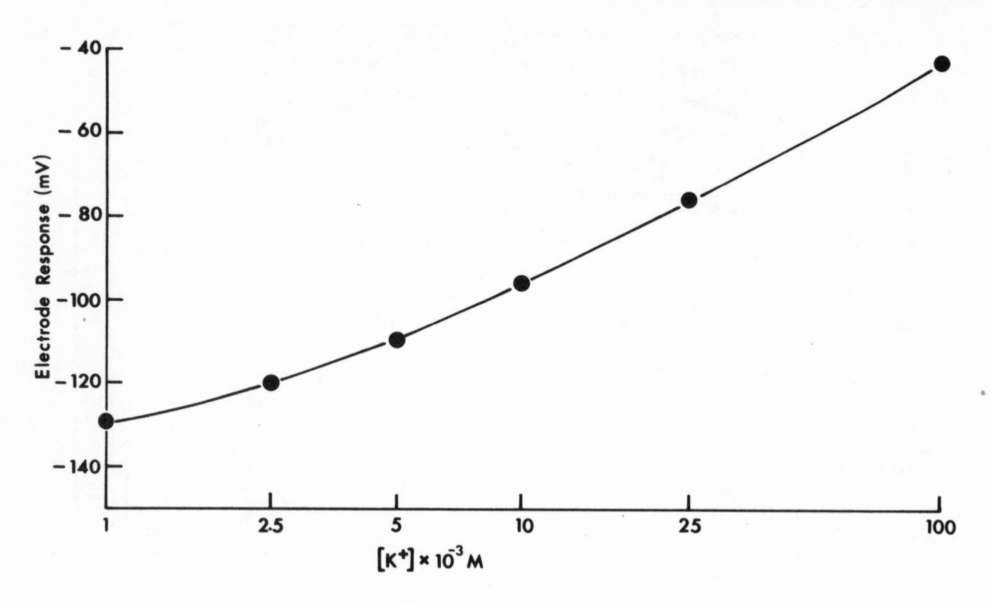

Fig. 3. Calibration curve. Voltage response of potassium electrode to various concentrations of KCl in Ringer solution.

stimulation of a dorsal root. The upper trace is the potassium potential (E_K) while the lower trace is the tissue potential (V) simultaneously recorded by the reference electrode. In Fig. 4B responses were recorded in the same experiment at the same location, using a combination of two reference electrodes. The potential change recorded between one of these electrodes and ground during stimulation was very similar to what had been seen before (cf. lower traces, A and B), but the response recorded differentially between the two reference electrodes (B, upper trace) differs markedly from that recorded previously between the K-sensitive and reference electrodes (A, upper trace) - it is not only of opposite polarity, but of very low amplitude. The traces of Fig. 4C show the common mode rejection capacity of the usual K electrode to 10 mV rectangular pulses applied between the animal and ground; again, the polarity of the differentially recorded potential was opposite to that of the potential with respect to ground, and the rejection ratio was 12:1.

In most of the experiments on the cerebral cortex (e. g., Fig. 6) changes in tissue O_2 concentration were monitored continuously from the surface of the brain with an oxygen electrode having a tip of about 0.5 mm (Fam, Nakhjavan, Sekely, & McGregor, 1966).

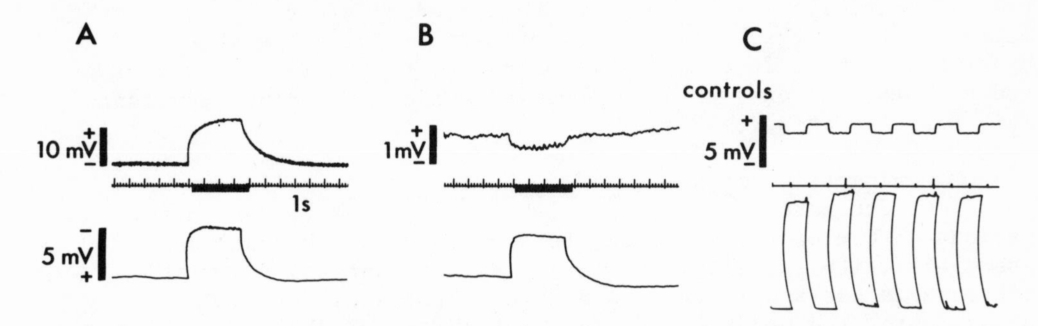

Fig. 4. Control tests of signal rejection capacity of K electrode, recording in cat spinal cord. Upper traces - with differential input from two micro-electrodes (note different gains); lower traces - recorded between single reference electrode and ground. (A) Differential response recorded by K-selective — reference electrode pair to tetanization of dorsal root (stimulus duration shown by black bar under time marks). (B) Responses recorded in same experiment as A under identical conditions, only using twin reference electrodes. (C) Responses of K-selective — reference electrode pair recorded during application of 10 mV pulses between ground and preparation.

OBSERVATIONS

1. On Neocortex In Guinea-Pigs

A. Changes in extracellular a_K during spreading depression. A liberation of cellular K^+ was proposed as the mechanism of cortical spreading depression by Grafstein in 1956 (cf. also Marshall, 1959). In agreement with this hypothesis we have confirmed the recent demonstration (Vyskočil, Kříž & Bureš, 1972) of a large increase in a_K in the neocortex during spreading depression. In the cortex of the guinea-pig, the characteristic slow negative wave evoked by applications of concentrated KCl (0.1 M) 1-2 cm away from the point of recording as in Fig. 5A or by a prolonged period of hypoxia (Fig. 5B), was associated with large increases of extracellular a_K - from resting levels of 2-5 mM to 30-50 mM. Tissue O_2 levels when recorded from the nearby brain surface were unchanged during the period of peak a_K. Changes in E_K and in tissue potential initially followed a very similar time course; the tissue potential commonly returned to the base-line more quickly, and often showed a clear undershoot, with only slow recovery (see Fig. 5B). This phase of relative positivity is evidently not due to a notable reduction of a_K below the initial base-line level. It is likely to be caused by electrogenic Na-K pumping - triggered by the excess of extracellular K^+- whose hyperpolarizing effect may well contribute to the later phase of reduced neuronal excitability observed during spreading depression (cf. Krnjević et al., 1966).

B. Changes in extracellular a_K during brief anoxia. Even a transient decrease in oxygen availability, caused by very brief periods of nitrogen inhalation (between 5 and 40 seconds), was associated with a small, but clearly significant and reversible increase in extracellular a_K. Fig. 6 illustrates such an experiment. The upper traces record E_K; the middle traces, the tissue potential with reference to ground; and the lowest traces show tissue oxygen levels, monitored at the brain surface. It can be seen that the inhalation of N_2 for periods (between two arrows) as short as 5-10 sec. caused a just detectable rise in a_K (Fig. 6A,B); the effects were markedly progressive with longer periods of N_2 inhalation (Fig. 6, C-F). However, recovery was very rapid, and in fact somewhat quicker than the restoration of the initial P_{O_2}. Even repeated brief anoxic episodes were not

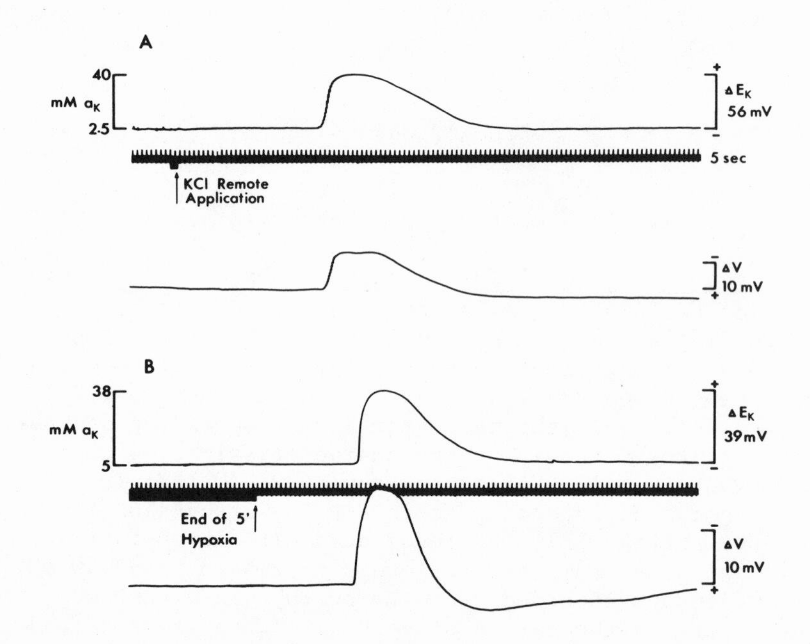

Fig. 5. Spreading depression in guinea-pig cortex
A. evoked by application of KCl (0.1 M) to cortex 1.5 cm away from recording electrode
B. occurring after a period of hypoxia
Upper traces - changes in potassium potential (ΔE_K)
Lower traces - changes in tissue potential (ΔV)

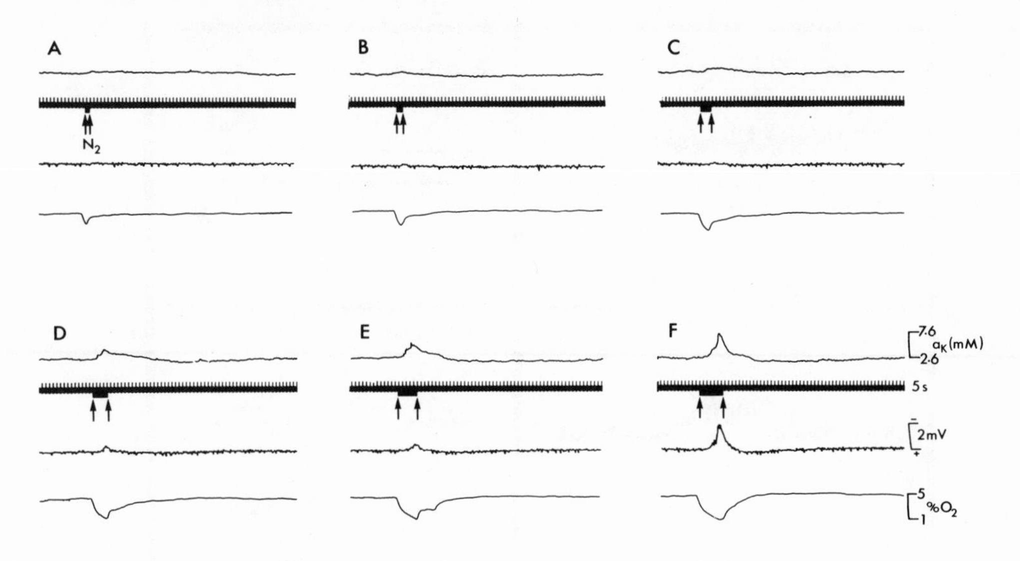

Fig. 6. Extracellular a_K changes in guinea-pig cortex during brief periods of anoxia.
Nitrogen inhalation during times between arrows — (A - 5 sec., B - 10 sec., C - 15 sec., D - 20 sec., E - 25 sec., F - 30 sec.). Upper traces, a_K; middle traces, ΔV - recorded at approximately 0.5 mm depth in the sensorimotor cortex; lower traces recorded by small oxygen electrode placed on cortical surface. (Reproduced in part from Fig. 4, In Morris, 1974).

associated with irreversible changes in a_K, although serious or prolonged hypoxia in association with cardiac or respiratory arrest not infrequently initiated "spreading depression" (cf. Fig. 5B). These observations are described elsewhere in greater detail (Morris, 1974).

2. On the Spinal Cord and Medulla of Cats

Barron and Matthews suggested in 1938 that ions released by afferent signals could, by accumulating around fine terminal branches of afferent fibres, cause the observed prolonged depolarization of these same or adjacent fibres. These slow potentials are not of purely academic interest, since they have been proposed as the mechanism of "presynaptic inhibition" (Eccles, 1964).

Many of the low threshold myelinated primary afferent fibres terminate in the dorsal horn of the spinal cord or have long central branches which ascend in the dorsal columns to make their first synapses in the cuneate and gracile nuclei of the medulla.

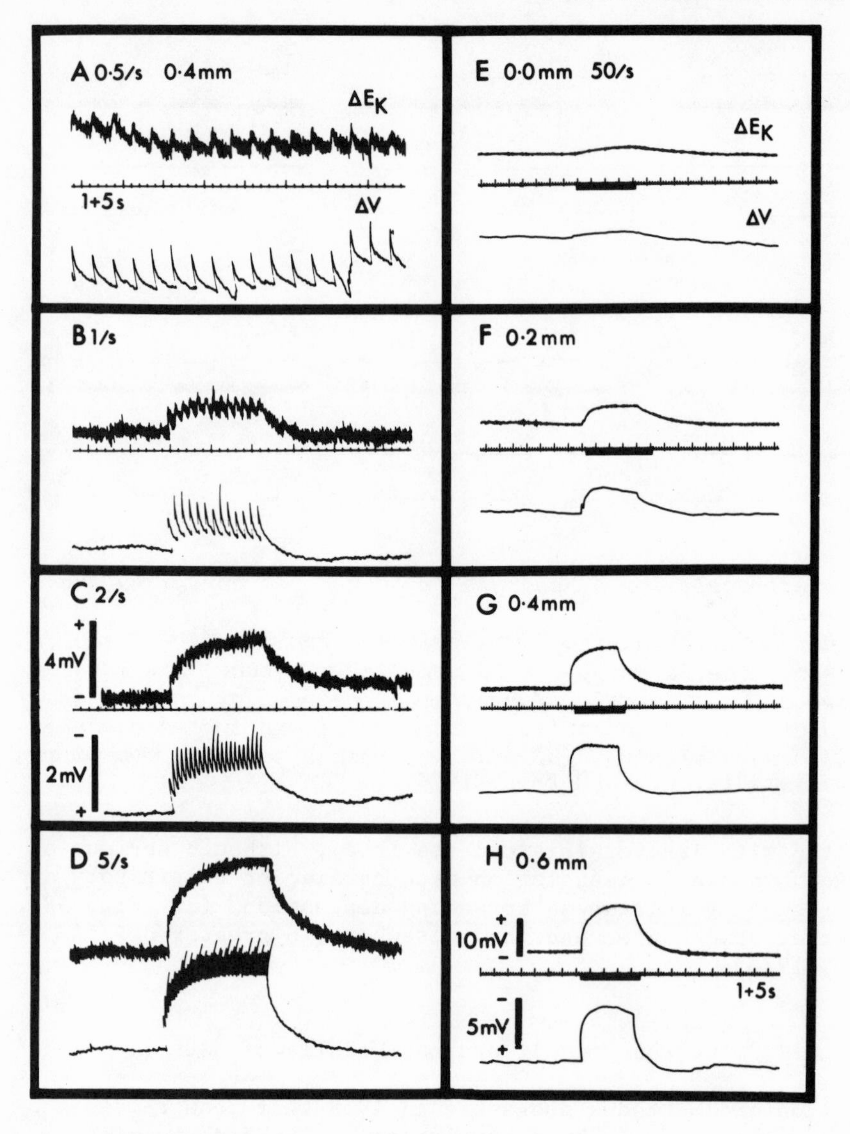

Fig. 7 Extracellular potassium activity in dorsal horn of cat spinal cord during afferent activity.
Upper traces - potassium potential (ΔE_K).
Lower traces - tissue potential (ΔV).
Stimulation of L7 dorsal rootlet (15V, 0.1 msec).
A - D: responses to varying stimulus frequency, at constant depth (0.4 mm)
E - H: responses recorded at different depths, during constant frequency of stimulation (50/sec).
(Reproduced in part from Fig. 2 in Krnjević and Morris, 1972).

In these densely packed synaptic regions - where the terminating fibres branch profusely and lose their myelin cover - there is a sharp increase in the axonal surface area/volume ratio: the release of K^+ and its accumulation extracellularly in association with axonal activity is necessarily likely to be very much greater per unit volume of tissue than in other areas (e. g., fibre bundles as in the dorsal columns).

A. Changes in extracellular a_K during afferent activity. A clear increase in extracellular a_K and in tissue negativity was evoked in the dorsal horn of the lumbar spinal cord by stimulation of a dorsal rootlet. As the K electrode penetrated to greater depths, into regions of more dense fibre branching and synaptic contacts, these potentials became larger (see Fig. 7, E-H). The time course of the change in a_K followed quite closely the change in tissue potential (ΔV) recorded at the same site; but the latter tended to fall off even during maintained tetanic stimulation, and usually showed a marked positive undershoot, probably indicating an electrogenic pumping process.

Even at very low rates of stimulation ($< 1/s$), definite

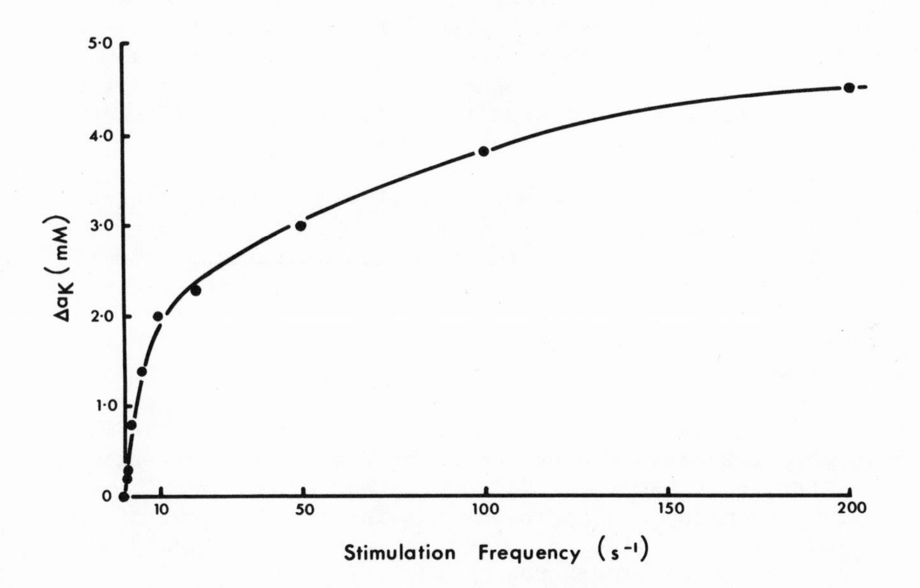

Fig. 8. Relation of a_K and stimulus frequency.
Ordinate - change in maximal a_K level, recorded at 0.4 mm depth in cat dorsal horn.
Abscissa - frequency of dorsal rootlet stimulation.
Data from experiment illustrated in traces of Fig. 7 (A - D).

small changes in a_K were recorded (Fig. 7A). The maximum amplitude of both ΔE_K and ΔV increased with the frequency of stimulation, as shown in Fig. 7(A-D). However, the rate of increase diminished at higher frequencies, so that the maximal changes in a_K showed a quite non-linear relationship to the stimulus frequency (Fig. 8); in fact Δa_K was approximately linearly related to the logarithm of the rate of stimulation. This feature may be caused by a progressively more rapid active reuptake of K^+, stimulated by the rise in extracellular a_K.

Similar observations were made in experiments on the cuneate nucleus. When a forelimb nerve was excited, the increase in extracellular a_K and the slow tissue potential changes (including the P wave recorded after a single volley) increased with depth; and they all reached a maximum at the same depth. The amplitude of these slow potential changes showed a high positive correlation ($r = 0.945$; $n = 45$) with changes in E_K, recorded simultaneously at various positions in or close to the cuneate nucleus.

These observations suggested that the slow tissue potentials in the presynaptic terminal regions may be caused by the extracellular accumulation of K^+ during afferent activity. However, the sensitivity of these terminals to raised levels of extracellular a_K is not known. Therefore in further experiments we artificially raised a_K in the cuneate nucleus by micro-injection of KCl or by superfusing the dorsal surface of the medulla with K-enriched solutions.

B. Effects of increasing extracellular a_K in the cuneate. Changes in presynaptic excitability and in the efficiency of synaptic transmission were measured by constant current stimulation via a third micro-electrode, attached near the ion-exchanger and reference electrodes. Direct stimulation of the afferent fibres near their terminals evokes antidromic responses in forelimb nerves (see R_1, Fig. 9) which provide an index of the number of fibres excited and thus of the excitability of the afferent terminals (Wall, 1958). At the same time, orthodromic trans-synaptic responses are evoked which can be recorded from the medial lemniscus (R_2, Fig. 9). These antidromic and orthodromic responses thus represent the input and output of the cuneate, and therefore give an indication of the efficiency of synaptic transmission through the nucleus.

When a_K was raised by the injection of a minute volume of 0.1 M KCl from a fourth micro-pipette, the integrated antidromic responses to varying intensities of cuneate stimulation were clearly increased (see lower traces, Fig. 10), presumably because the terminals were significantly depolarized.

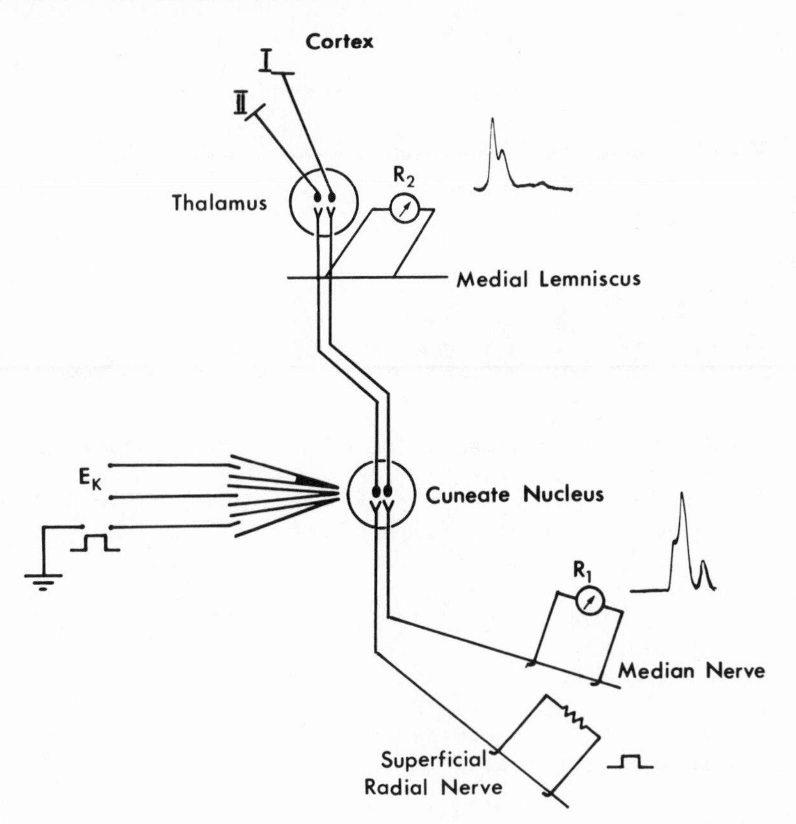

Fig. 9. Diagram of stimulating and recording arrangements in experiments where extracellular a_K *was artificially increased.* R_1 *- antidromic, and* R_2 *- orthodromic responses evoked by stimulation of afferent fibre terminals in the cuneate nucleus, via third micro-electrode attached to K electrode.*

At the same time, the trans-synaptic responses were either unchanged or increased to a lesser extent than the antidromic responses, so that the slopes of input-output curves were reduced: the efficiency of synaptic transmission was therefore lower.

These observations show that the afferent terminals are sensitive to changes in external a_K and that a rise in a_K is likely to reduce (rather than increase, cf. Cooke & Quastel, 1973) the efficiency of transmission through cuneate synapses. However, we have not enough information to state that the increases in a_K observed near the presynaptic terminals during

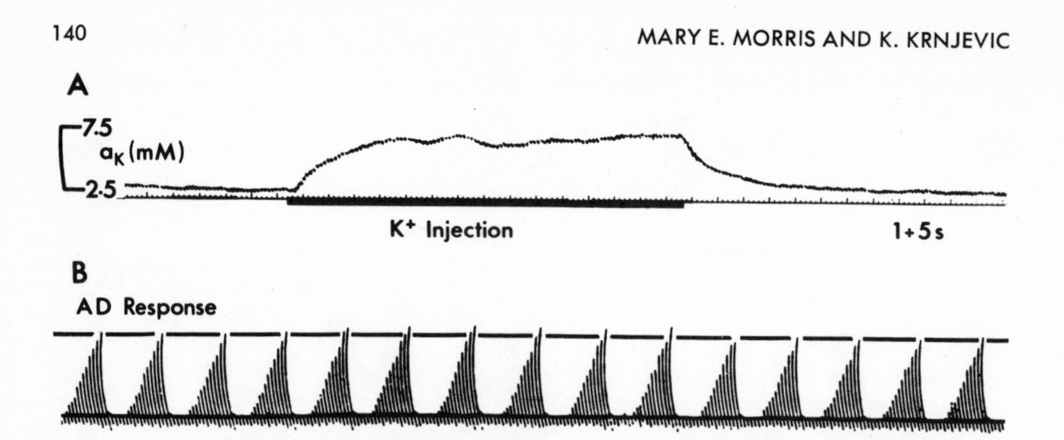

Fig. 10. Effect of increased extracellular a_K on excitability of afferent terminals.
A. Potassium activity recorded in cuneate nucleus, depth 0.6 mm
B. Integrated antidromic potentials recorded from peripheral forelimb nerve in response to cycles of automatically incremented constant current stimulation of fibre terminals via third micro-electrode, attached to K electrode. KCl (0.1 M) injected from fourth micro-pipette, 100 µm away, during time marked by horizontal black bar.

afferent activity are fully sufficient to account for all the known changes in synaptic function.

ACKNOWLEDGEMENTS

It is a pleasure to thank the Medical Research Council of Canada for financial support, and Joshua Wu and Edna Leech for their technical assistance.

REFERENCES

AMES, A., III, HIGASHI, K. and NESBETT, Frances B. 1965. Relation of potassium concentration in choroid-plexus fluid to that in plasma. *J. Physiol. 181,* 506-515.

BARRON, D. H., and MATTHEWS, B. H. C. 1938. The interpretation of potential changes in the spinal cord. *J. Physiol. 92,* 276-321.

COOKE, J. D. and QUASTEL, D. M. J. 1973. The specific effect of potassium on transmitter release by motor nerve terminals and its inhibition by calcium. *J. Physiol. 228,* 435-458.

DUBNER, H. H. and GERARD, R. W. 1939. Factors controlling brain potentials in the cat. *J. Neurophysiol. 2,* 142-152.

ECCLES, J. C. 1964. The Physiology of Synapses. Springer-Verlag; Berlin, Germany, p.316.

FAM, W. M., NAKHJAVAN, F. K., SEKELY, P. and McGREGOR, M. 1966. *Proc. Int. Symp. Cardiovasc. & Respir. Effects of Hypoxia, Kingston, Ont. 1965*, pp. 375-390.

GRAFSTEIN, B. 1956. Mechanism of spreading cortical depression. *J. Neurophysiol. 19*, 154-171.

HINKE, J. A. M. 1961. The measurement of sodium and potassium activities in the squid axon by means of cation-selective glass micro-electrodes. *J. Physiol. 156*, 314-335.

KRNJEVIĆ, K. and MORRIS, M. E. 1972, Extracellular K^+ activity and slow potential changes in spinal cord and medulla. *Can. J. Physiol. Pharmacol. 50*, 1214-1217.

KRNJEVIĆ, K., RANDIĆ, M. and STRAUGHAN, D. W. 1966. An inhibitory process in the cerebral cortex. *J. Physiol. 184*, 16-48.

KRNJEVIĆ, K. and SCHWARTZ, S. 1967. The action of γ-aminobutyric acid on cortical neurones. *Expl. Brain Res. 3*, 320-336.

KUFFLER, S. W. 1967. Neuroglial cells: Physiological properties and a potassium mediated effect of neuronal activity on the glial membrane potential. *Proc. Roy. Soc. Ser. B 168*, 1-21.

MARSHALL, W. H. 1959. Spreading cortical depression of Leão. *Physiol. Rev. 39*, 239-279.

MORRIS, M. E. 1974. Hypoxia and extracellular potassium activity in the guinea-pig cortex. *Can. J. Physiol. Pharmacol.* 52 (in press).

VYSKOČIL, F., KŘÍŽ, N., and BUREŠ, J. 1972. Potassium-selective microelectrodes used for measuring the extracellular brain potassium during spreading depression and anoxic depolarization in rats. *Brain Research 39*, 255-259.

WALKER, J. L. 1971. Ion specific liquid ion exchanger microelectrodes. *Analyt. Chem. 43*, 89A-93A.

WALL, P. D. 1958. Excitability changes in afferent fibre terminations and their relation to slow potentials. *J. Physiol. 218*, 671-689.

DISCUSSION

Questions: (Brown)

1. What is the voltage sensed by the ion selective electrode?
2. What is time constant of the 2 recording electrodes?
3. What is the space from which you are recording?

Answers: (Morris)

1. The voltage is the same as that seen by the reference electrode (10-50μm away), plus the local E_K.
2. The time constant of the electrode recording system is <1 sec.
3. Both electrodes are recording from the extracellular space as shown by the absence of any standing negative potential, low values of E_K and low resistance to ground.

Question: (McDonald)

Since neurotransmitters of synapses (e.g. cholinergic, adrenergic, aminergic synapses) are released presynaptically onto the postsynaptic membrane, it seems that iontophoresis of K^+ onto the synaptic cleft region is an effect on transmission and not the initial cause of transmission.

Answer: (Morris)

Agreed: We are not suggesting that K^+ is the synaptic transmitter. Our interest is in the effect of increases in extracellular K^+ (caused by K^+ release from active fibres) on the efficiency of synaptic transmission presumably through changes in transmitter release.

Questions: (Zeuthen)

1. How long did it take for the extracellular K+ to regain its original value during spreading depression?
2. Did you measure O_2 uptake in the spreading depression?

Answers: (Morris)

1. Several minutes.
2. No.

Question: (Meyers, NINDS)

Why should restoration of oxygen following a period of hypoxia initiate a wave of spreading depression as suggested by one of your slides? Or was the spreading depression induced by the hypoxia itself?

Answer: (Morris)

It may be that different types of neuronal tissues are affected differently and that during recovery from hypoxia - the uptake of potassium, which is released as you saw in one of the experiments, may be greater in some cells than others with the production of sinks and sources and current flow. Also, hypoxia itself can cause spreading depression.

Question: (Berman)

Do you consider variation in extracellular K^+ to be a regulator of activity of peripheral and central neuronal activity? Would this regulation be selective with respect to localization or topography?

Answer: (Morris)

I think that changes in extracellular potassium activity such as those we have demonstrated are involved in physiological neuronal activity. Yes, I suppose that regulation would be different in different regions.

MICROELECTRODE RECORDING OF ION ACTIVITY IN BRAIN

Thomas Zeuthen*, R.C. Hiam and I.A. Silver

Dept. of Pathology, University of Bristol

Great Britain

INTRODUCTION

This paper deals with some technical aspects of the registration of extracellular potassium (K^+) activity in the cortex of rats during hypoxia either by means of single-barrelled microelectrodes utilizing K^+-sensitive liquid ion-exchanger (Corning Catalog no. 477317) (Cornwall et al., 1970; Khuri, 1971; Walker, 1971, 1973) or by means of double-barrelled microelectrodes, where one barrel measures electrical potential (the DC-barrel) and the other barrel measures electrical potential plus a potential determined by the K^+-activity. (Vyskocil et al., 1972). We used a different method of producing double-barrelled electrodes from that described by Khuri et al., (1972a; 1972b) in order to obtain long shanks (5-10mm).

A few measurements were performed with K^+-sensitive microelectrodes of the all-glass construction devised by Thomas (1970, 1973).

The liquid ion-exchanger microelectrode consists in principle of a micropipette in which the internal wall of the tip-region is made organophilic-hydrophobic so that a column of the organic ion exchanger remains held there, when the rest of the inside of the electrode is filled with a conducting aqueous solution and the tip of the electrode is in tissue or a solution. The

---

* T. Zeuthen is from the Institute of Neurophysiology, University of Copenhagen, Denmark.

performance of the liquid ion-exchanger in respect to sensitivity and selectivity has been described in detail by Sandblom, et al., (1967 a, b). Walker (1971) gives an empirical equation:

$$E = E_o + \frac{nRT}{F} \log_e \left(a_i + K_{ij} a_j \frac{Z_i}{Z_j}\right) \qquad (1)$$

E is the recorded electrical potential in volts; E_o is a potential determined by the choice of reference-electrode; R is the gas constant (8.3 joules deg^{-1} mol^{-1}); T is the temperature (^oK); F is the Faraday number (96,500 coulomb $equivalent^{-1}$); n is an empirical constant chosen so that the slope of E versus $\log_e (a_i)$ is $\frac{nRT}{F}$ ($K_{ij}a_j = 0$); a_i is the activity of the ion we are interested in (valence Z_i); a_j is the activity of any interfering ions (the valence Z_j should have the same sign as Z_i). K_{ij} is the selectivity constant. In this paper only the interference from Na^+ ($Z_i = Z_j = 1$) will be considered.

The single-barrelled K^+-sensitive ion-exchanger microelectrode has previously been used intracellularly in giant Aplysia neurons (Kunze and Brown, 1971; Walker, 1971, 1973), in muscle fibers (Khuri, 1971) and intra-luminally in rat kidney proximal tubules (Khuri et al., 1971). The double-barrelled electrode has been used intracellularly in cells of the distal tubule (Khuri et al.,, 1972a) and in muscle fibers (Khuri et al., 1972b).

METHODS

Rats, Surgical Procedure

White Wistar rats (160g-500g, average 250g) were anaesthetized with 25% w/v urethane (ethyl carbamate, B.D.H., Poole, Dorset) at a rate of 0.6 ml/100g body weight, given by intraperitoneal injection. 0.2 ml local anaesthetic Lignocaine HCL (Xylocaine, Astra Ltd.) was administered subcutaneously around each external auditory meatus, along the top of the head and in the throat region. A polyethylene cannula (diam. 3mm) was introduced into and secured by a ligature, to the central end of the incised trachea. The vertical part of both the external ear canals was cut and the rat placed in a stereotaxic head holder that was electrically isolated from the rest of the set-up, and placed inside a Faraday cage. The rat was supported on a foam rubber pad, the skin over the upper surface of the head was excised and the periosteum of the skull removed to expose the bone. Two holes (diameter 1mm) were drilled 4mm in front of the Bregma suture and 3mm to the left and right sides of the midline. An E.E.G. electrode was inserted into one hole and a macro-oxygen electrode (leading-off area about $0.01mm^2$) in the other. Both were fixed with "Simplex" acrylic dental cement (Dental Fillings, Ltd.). For the microelectrode recordings a larger hole (diameter 3mm) was drilled 2mm behind the Bregma suture and 3mm from the midline; the dura was removed and the exposed cortex covered with 3% agar (Noble Agar, Oxoid Ltd.) in phosphate buffered saline (Dulbecco A, Oxoid Ltd.).

Up to four microelectrodes could be advanced into the brain through this hole by means of four separate micromanipulators (Prior, Ltd.) rigidly attached to the sterotaxic holder. The microelectrodes were fixed to the micromanipulators by means of teflon holders.

The electrocardiogram (E.C.G.) was recorded from two needle electrodes, one in the right fore-limb and the other in the left hind limb. The respiration was recorded by means of an airtight plastic bag placed between the rat and the foam rubber pad; the interior of the bag was connected to a Grass volumetric transducer (Model PT5). Respiratory gases were administered from a Boyles apparatus (British Oxygen Co. Ltd.) via a metal tube (diameter 5mm) the end of which was positioned within 5mm of the outer opening of the tracheal cannula. Hypoxia was produced with a gas mixture of 5% v/v carbon dioxide in nitrogen or nitrous oxide.

Two types of reference electrode were used; a: an 0.3mm diameter silver wire 2 cm long, chlorided in 0.1M sodium chloride by a voltage of about 3 volts for 30 secs. b: a similarly chlorided silver wire inserted into one end of a 5cm long polyethylene tube (outer diameter 2mm, inner diameter 1mm) filled with 2M KCl that was gelled by means of 3% agar which gave a solidified salt bridge connection between the rat and the silver-chloride. Both electrodes were inserted under the rat's skin near the ear and served as a reference for both the oxygen and ion sensitive electrodes.

All recordings were monitored on a Grass polygraph (Model 5D).

Microelectrodes

Single-barrelled liquid ion-exchanger electrodes were prepared as described by Walker (1971, 1973). Micropipettes were pulled from Corning 7740 glass with a tip diameter not less than 0.5μm (determined by electron microscopy). Tips smaller than 0.5μm diameter tended to clog in the next procedure when they were dipped into 1% v/v Siliclad (Clay Adams, Inc., N.Y.) and subsequently baked at 200°C for 1 hour to render them water repellent. The siliconized tips were dipped for about ten seconds in Corning 477317 liquid ion exchanger of which they took up a column about 200μm long. The electrodes were filled with 2M potassium chloride by local heating (Zeuthen, 1971). A 2cm length of 0.3mm diameter silver wire chlorided for 30 sec in 0.1M NaCl at 2 volt relative to a bright silver wire, was used as an internal reference and fixed in the opening of the pipette shaft by adhesive sealant silicone rubber (General Electric); this also prevented evaporation of the electrolyte. Electrodes losing more than 10μm of ion exchanger within the first 15 minutes after filling with KCl were discarded since such electrodes tended subsequently to lose their exchanger completely.

Double-barrelled microelectrodes were made from glass (length 7.5cm; inner diameter, 1.2mm; outer diameter 1.8mm) washed in ethanol and while still wet placed vertically in nitric acid. After the formation of NO_2 vapour had ceased the tubes were rinsed 10 times in distilled water and dried. Two pieces of glass with their ends offset about 0.5cm were held together in the two chucks of a vertical gravity puller (Frank and Becker, 1964). The glass was heated around the middle and when soft enough one chuck was rotated 180° in respect to the other. Thereafter the two holders were pulled apart by gravity which produced two double-barrelled electrodes each with one shaft 0.5cm longer than the other. We readily obtained shank lengths of 5mm to 10mm, a taper at the tip region of about 1/10 and a tip diameter of about 0.3μm per barrel as seen from electron microscopy. Since Siliclad solutions (0.1% to 2% v/v in 1-chloronaphthalene were tried) clogged tips of this size the glass was made hydrophobic by exposing it to vapour from dimethyl-dichloro-silane (DDS) (Hunyar, 1959). A 100ml beaker was filled with 25ml DDS and closed with a metal lid; the long shaft of the double-barrelled electrode was pushed through a 2mm diameter hole drilled in the lid and sealed into it with modelling clay. In this way the interior of that barrel was exposed for 60 sec to the vapour inside the beaker, whereas the other barrel was not. Since DDS is very volatile, the beaker was closed with a glass lid sealed with silicone grease when not in use. The electrodes were then baked for 1 hour at 100°C. The unexposed barrel was filled with distilled water down to about 2mm from the tip by means of a capillary tube pulled to a 5cm long shank with a diameter of about 75-100μm, attached to a syringe. The shank of the hydrophobic barrel was filled in a similar manner with a 5mm long column of liquid ion exchanger (Corning code 477317 potassium exchanger) (Walker, personal communication). For about one half of the electrodes the tips would fill completely by capillary action within 10-20 min. The other electrodes contained air bubbles in one or both shanks; these were removed by means of a 25μm platinum 80%, rhodium 20% wire heated by an electric current and brought up to the outside of the glass at the air-water or air-exchanger interface furthest away from the tip. This caused the water (or exchanger) to distill towards the tip and the bubble to move towards the shank. The shaft of the barrel containing the exchanger was then filled with 0.5M KCl and the other barrel with 2M KCl. The DC-barrel was finally allowed to equilibrate by placing the tip of the electrode in 0.5M KCl. As with the single-barrelled electrodes a chlorided silver wire served as an internal reference electrode.

In another procedure for making double-barrelled potassium-sensitive microelectrode the DC-barrel was first filled with water immediately after pulling. Then the whole electrode was exposed to the DDS vapour in a 5 litre beaker. No baking was performed afterwards. The empty barrel was subsequently filled with the exchanger as described above; heating with the wire was always necessary as the tip was plugged with water. This method produced an electrode where the first 50-150μm of the 'ion exchanger' shank near the tip was filled with water. This water column could be removed or shortened if required by placing the heating wire 50μm from the tip.

Single-barrelled potassium sensitive all-glass microelectrodes were constructed in the same way as for the all glass sodium sensitive microelectrode devised by Thomas (1970, 1973). The potassium sensitive tip was made from Corning Glass NAS 27-4. The electrodes were filled by local heating (Zeuthen, 1971) with 2M potassium chloride and allowed to hydrate overnight or longer with the tips immersed in distilled water.

Electronics

The internal reference electrode of the ion sensitive microelectrode (or ion sensitive barrel) was connected to the input of an electrometer (Analog devices 311K coupled as a unity gain follower); the corresponding electrical potential-recording microelectrode (or DC-barrel) was connected to a low-input-current operational amplifier (Analog Devices 40K), coupled as unity gain follower. These electrometers were placed in an aluminium box 10cm above the rat in order to keep the input leads as short as possible. The outputs of the electrometers were fed via a multiway screened cable (through which the power supply for the electrometer was supplied as well) each to the positive input of a differential amplifier, the negative input of which was connected to a constant but adjustable voltage. Thus any resting potential (E in equation 1 when a_i and a_j are constant) read by the electrodes could be adjusted to zero at the outputs of the differential amplifiers and only the deviations from this resting value were amplified either 1, 10 or 100 times and filtered by an upper limiting frequency of 1.6 HZ in order to reduce high frequency noise and to minimize the effect of electrostatic interference, e.g. from moving objects.

In order to obtain the difference between the potential recorded with the K^+-sensitive barrel and the potential recorded with the DC-barrel, the positive input of another differential amplifier was fed by the amplified signal from the K^+-sensitive barrel and the negative input was fed by the signal from the corresponding potential-sensitive barrel.

All differential amplifiers were enclosed in an aluminium box which was connected via the shield of the multiway cable to the box containing the electrometers.

RESULTS

a) Properties of the double-barrelled electrode in KCl solutions.

The electrodes were treated in 0.1M KCl, in 0.01 M KCl and in 0.01M KCl + 0.1M NaCl. The sensitivity (the change in potential for a ten-fold change in activity) of the K^+-sensitive barrels ($\frac{nRT}{F}$ of eq. 1) was constant for at least 8 hours and had values between 40mV and 56mV. The selectivity constant (K_{ij} of eq. 1) was about 0.01 in respect to Na^+. The time constant of those barrels that were completely filled with exchanger was less than 10 sec. The potential of the DC-barrel (tip potential plus liquid junction potential) varied less than 4mV for about 75% of the electrodes when tested in 0.1M KCl, 0.01 M KCl + 0.1NaCl and in phosphate buffered saline (Dulbecco A, Oxoid Ltd.).

The impedances of the two barrels were measured with a Keithley electrometer (610C) in 2M KCl and in 0.1M KCl. In 2M KCl the impedance of the barrel when the tip was completely filled with exchanger was 2000 times that of the DC-barrel which varied between 1MΩ and 20MΩ. This agrees well with the values obtained by Khuri (1972b) and with the value of the specific resistivity of the K^+-sensitive ion-exchanger (11.3·10Ωcm) compared to the specific resistivity for 2M KCl (5.5Ωcm); the value was found from the resistance through a capillary tube filled with a column of ion-exchanger with a column of 2M KCl on each side. The impedance of the ion-exchanger barrel rose 10-100% when measured in 0.1M KCl compared to the impedance in 2M KCl. The impedance of ion-exchanger barrels with a watery solution in the tips was much lower; if the column of solution was 100μm the impedance was typically 100MΩ. Stability and low frequency noise (<2HZ) of the potential was measured in 2M KCl and phosphate buffered saline (Dulbecco A, Oxoid Ltd.). The potential recorded was stable within 1mV for at least 8 hours in both solutions, for both the DC-barrel and the exchanger barrel; but the potentials recorded with the exchanger barrel had a low frequency noise (about 0.1HZ) with an amplitude of about 0.3mV in the 2M KCl and about 1.0mV in the saline.

b) Results in the brain.

Figure 1 shows a recording with a double-barrelled microelectrode (sensitivity 45mV) 1mm below the surface of the cortex. The potential obtained with the potential-sensitive barrel is called DC in the figure and the potential recorded with the ion-sensitive barrel is called pK + DC. The difference between these two potentials (called pK in the figure) is obtained with the differential amplifier and represents the potential that arises from the K^+-activity. If electrodes were selected so that the tip potential of the reference barrel varied less than 2mV between 150mM KCl and 150 mM NaCl, and if we assume that the tip potential is the same in these solutions and in the brain then the resting level of K^+ in the brain corresponds to 3mM. This value was also obtained in the ventricle. The E.E.G. and E.C.G. are also shown. Before the rat was made anoxic at a the electrical potential was stable but there are slow variations in pK (frequency of about 0.001 HZ, amplitude 5mV). These variations disappered after the rat had been dead for a few minutes (heart stopped completely). A potential synchronous with the breathing was picked up equally well with the potential and the K^+-sensitive barrel so that it cancelled out in the pK trace. The two small peaks, one at a and one 1 minute before, show an electrostatic effect caused by the jet of gas mixture as it passed the head of the rat; the first peak was caused by gas not directed at the tracheal cannula; the second represents the onrush of gas that caused the anoxia. The immediate effect of the nitrous oxide + 5% carbon dioxide (start at a) is a decrease in pK of about 3mV (corresponding to a decrease in K^+-activity of 16%). This decrease lasts for about 40 sec. and is followed by an increase in pK of about 5mV (which corresponds to an increase in K^+-activity of about 29%). When breathing stopped at b, there is a decrease in E.E.G. and a slowing down of heart rate. After about 5 min. this is followed by a big increase in pK of about 30mV (a 4.6 times increase in K^+-activity). Similarly the D.C. potential showed a decrease at the onset of anoxia but smaller, and in this recording starting about 5 sec. later than that obtained from the exchanger barrel; this time difference varied from place to place in the brain. Recordings performed with one single-barrelled K^+-sensitive microelectrode plus one single barrelled potential measuring electrode, tips 1mm apart, 1mm below the surface of the cortex showed a similar pattern. Two experiments were performed with the tip of an all-glass K^+-sensitive electrode 1mm away from the tip of a single-barrelled exchanger-electrode; the two recordings were similar allowing for differences in sensitivity and time constants. The tissue PO_2 as measured with the platinum macroelectrode fell during the decrease in pK; it reached its minimum value when the pK turned positive.

DISCUSSION

We have developed a reliable method of producing a straight double-barrelled microelectrode where one barrel records the change in K^+-activity plus electrical potential and the other change in

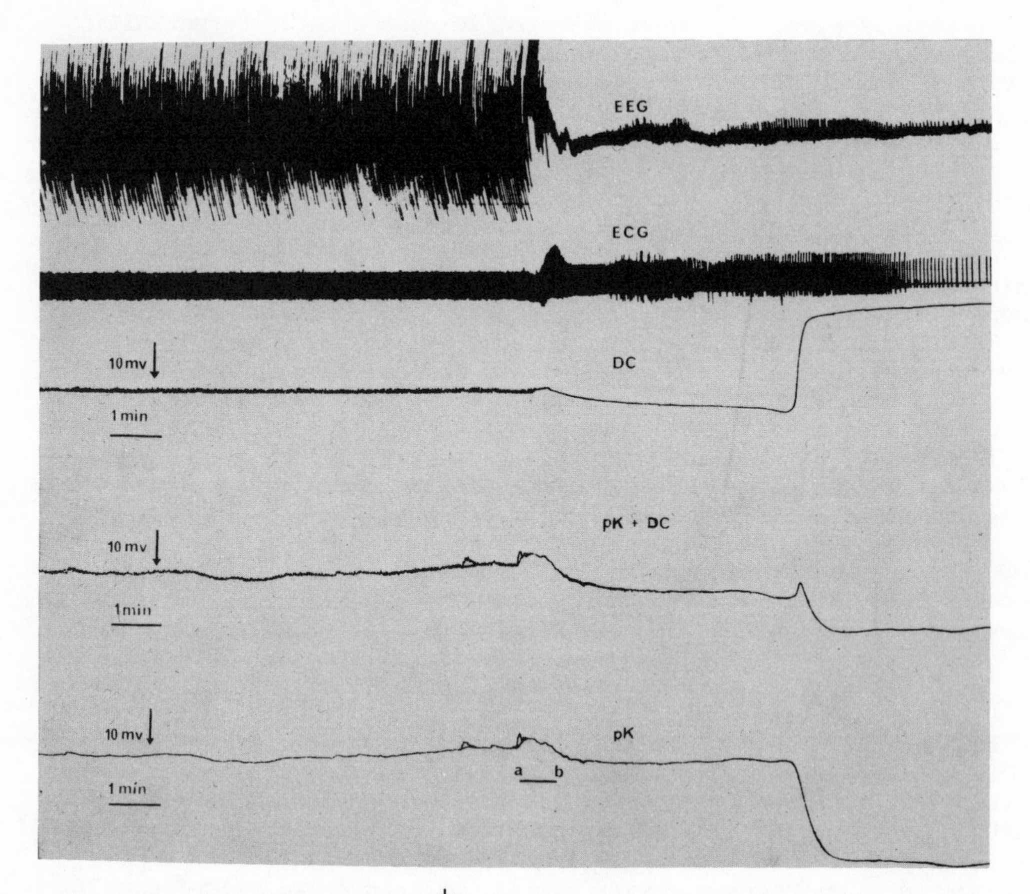

Figure 1. The response in K^+-activity (pK) and electrical potential (DC) as recorded during an anoxic incident by means of a double-barrelled potassium sensitive microelectrode. a indicates the onset of hypoxia, b the stopping of breathing. For further explanation see text.

potential only. The small diameter of the tips of the electrodes (about 0.3μm per barrel) and their long shanks make them suitable for measuring quite deep in brain tissue. Recordings with this type of electrode in KCl-solutions of physiological concentrations showed similar properties to those of existing types of K^+-sensitive microelectrodes (Walker, 1971, 1973; Khuri, 1971; Khuri et al., 1971; Khuri et al., 1972a, 1972b). The method of making the glass hydrophobic by means of dichlorodimethyl silane solved the problem of clogged tips which occurred with the method of Walker (1971). The problem with the liquid ion-exchanger leaking out of the electrode tip (Walker, 1971) was solved by Silicone coating the whole electrode and having water in the tip for applications where large time constants are no problem. Furthermore this improved the insulating properties of the electrode (Engbak and Guld, 1971).

The electrode recorded the correct resting level of K^+-activity in the cortex as well as in the ventricle (3mM) if it was assumed that the tip potential of the reference barrel was the same in the brain as in the 150 mM NaCl. This suggests that neither the tissue matrix nor the composition of the extracellular fluid influences the tip potential significantly.

The recordings in the cortex of the rat during anoxia produced similar results to those obtained with one single-barrelled electrode that recorded electrical potential plus one single-barrelled liquid ion-exchanger or all-glass K^+-sensitive electrode in terms of an initial decrease in extracellular K^+, which probably results from a hyperpolarization of the brain cells (Kolmodin and Skogland, 1959; Grossman and Williams, 1971). This hyperpolarization was not found if the animals were paralyzed with Flaxedil. The subsequent potassium rise probably results from release of K^+ from the cells (Bradbury, 1971; Fenn and Gerschmann, 1949). The final large increase in extracellular potassium and simultaneous negative shift in potential are probably due to anoxic depression (Leao., 1944; Vyskocil et al., 1972). But the fact that there exists a time difference between the onset of the decrease in electrical potential and the onset of decrease in K^+ and that this time difference varies spatially, implies that the double-barrelled electrode is necessary for the investigation of ion movements during brain hypoxia.

Acknowledgement

We would like to thank Professor Britton Chance at Philadelphia for facilities, advice and encouragement. This work was supported by U.S.P.H.S. Grant No. 1 P01 NS 1 0, 939-01 NSP-A.

Note added in proof

The method for producing the double-barrelled K^+-sensitive microelectrode can be used to produce Cl^--sensitive electrodes as well, if Cl^--ion exchanger is used.

REFERENCES

Bradbury, M.W.B. Potassium Homeostasis in Cerebrospinal Fluid. p. 138 in Alfred Benzon Symposium III. Eds. Siesjo, B.K. and Sorenson, S.C. Munksgaard, Copenhagen, 1971.

Cornwall, M.C., Peterson, D.F., Kunze, D.L., Walker, J.L. and Brown, A.M. Intracellular potassium and chloride activities measured with liquid ion exchanger microelectrodes. Brain Res. 1970, 23, 433.

Engbak, E. and Guld, C. in Proceedings of the Second Nordic Meeting on Medical and Biological Engineering. June, 16-19, 1971, Oslo.

Fenn, W.O. and Gerschmann, W. The Loss of Potassium from Frog Nerves in Anoxia and other Conditions. J. Gen. Physiol. 1949, 33, 195.

Frank, K. and Becker, M.C. Microelectrodes for Recording and Stimulation. In "Physical techniques in biological research" Vol. V, pp. 23-89. Ed. Nastuk, W.L., Academic Press, New York, 1964.

Grossman, R.G. and Williams, V.F. Electrical Activity and Ultrastructure of Cortical Neurones and Synapses in Ischaemia. In "Brain Hypoxia", p. 61. Eds. Brierly, J.B. and Meldrum, B.S. S.I.M.P./William Heineman. S.I.M.P./Medical Books Ltd., London, 1971.

Hunyar, A., "Chemie der Silicone", Verlag Technik, Berlin, 1959.

Kolmodin, G.M. and Skogland, C.R. Influence of Asphyxia on Membrane Potential Level and Action Potentials of Spinal Moto- and Inter-neurones. Acta Physiol. Scand., 1959, 45, 1.

Khuri, R.N. Intracellular Potassium and the Electrochemical Properties of Striated Muscle Fibers (Abstract). Proc. Intern. Union Physiol. Sci., 1971, 9, 301.

Khuri, R.N., Aguilian, S.K. and Wise, W.M. Potassium in the rat kidney proximal tubules *in situ*: determination by K^+-selective liquid ion-exchanger microelectrodes. *Pflug. Arch. Ges. Physiol*. 1971, *322*, 39.

Khuri, R.N., Aguilian, S.K. and Kalloghilan, A. Intracellular Potassium in Cells of the Distal Tubule. *Pflug. Arch. Ges. Physiol*. 1972a, *335*, 297.

Khuri, R.N., Hagfar, J.J. and Aguilian, S.K. Measurement of intracellular potassium with liquid ion-exchange microelectrodes. *J. Appl. Physiol*., 1972b, *32*, 419.

Kunze, D.L. and Brown, A.M. Internal Potassium and Chloride activities and the effects of acetylcholine on identifiable Aplysia neurones. *Nature*, 1971, *229*, 229.

Leao, A.A.P. Spreading depression of activity in the cerebral cortex, *J. Neurophysiol*. 1944, *7*, 359.

Sandblom, J.P., Eisenman, G. and Walker, J.L., Jr. Electrical Phenomenon associated with the Transport of Ions and Ion Pairs in Liquid Ion Exchanger Membranes. I. Zero Current Properties. II. Non-Zero Current Methods. *J. Phys. Chem*. 1967*a*, *b*, *71*, 3862 and 3871.

Thomas, R.C. New Design for Sodium-Sensitive Glass Microelectrodes. *J. Physiol*. 1970, *210*, 82P.

Thomas, R.C. pNa Microelectrodes with Sensitive Glass Inside the Tip. In "Ion Selective Microelectrode". Ed Hebert, N.C. and Khuri, R., Dekker, New York (1973) in Press.

Walker, J.L., Jr. Ion Specific Liquid Ion Exchanger Microelectrodes. *Anal. Chem*., 1971, *43*, 89A.

Walker, J.L., Jr. Liquid Ion-Exchanger Microelectrodes for Ca^{2+}, Cl^- and K^+. In "Ion Selective Microelectrodes" eds. Hebert, N.C. and Khuri, R.N., Dekker, New York, 1973 (in Press).

Vyskocil, F., Kriz, N. and Bures, J. Potassium-selective microelectrodes used for measuring the extracellular brain potassium during spreading depression and anoxic depolarization in rats. *Brain Research*. 1972, *39*, 255.

Zeuthen, T., A Method to Fill Glass Microelectrodes by Local Heating. *Acta physiol. Scand*. 1971, *81*, 141.

DISCUSSION

Question: (Krnjevic)

Could the apparent fall in a_K at the start of hypoxia be an artefact? We have never seen such a change in our experiments; and it seems very hard to understand why there should be a fall in extracellular K^+ under these conditions!

Answer: (Zeuthen)

That the hyperpolarization might be an artefact can't be excluded on grounds of the ion-exchange electrode data available at present; but on the other hand, hyperpolarization as an initial response to hypoxia has been observed with other methods, e.g. Grossman, R.G. and Williams, In Brain Hypoxia, Eds. Brierly, J.B. and Meldrum, B.S. S.I.M.P./William Heineman. S.I.M.P./Medical Books Ltd., London, 1971. However the hyperpolarization is not present if anaesthetised animals are paralysed with Flaxedil.

Question: (McDonald)

Did you use theta glass to make double-barrel glass pipettes?

Answer: (Zeuthen)

We used Corning 7740 of a round cross section.

Questions: (Morris)

1. Perhaps I missed something you said - how did you produce anoxia?
2. In some of your experiments the animal stopped breathing - which is a different situation from pure hypoxia?

Answers: (Zeuthen)

1. Nitrogen and 5% CO_2, or Nitrous oxide and 5% CO_2 was supplied via a tracheal canula.
2. Yes.

V. Myocardium

INTRACELLULAR POTASSIUM AND CHLORIDE ACTIVITIES IN FROG HEART MUSCLE

John L. Walker and Roger O. Ladle
Department of Physiology
College of Medicine
University of Utah
Salt Lake City, Utah 84112

We have been using K^+ and Cl^- liquid ion exchanger microelectrodes to measure intracellular activities of K^+ and Cl^- in frog heart. In these studies we have used three kinds of heart muscle: ventricle, atrium and sinus venosus. The preparations are small pieces of tissue cut from the heart and pinned down in a chamber where they are superfused with frog Ringers solution. The ventricular and atrial preparations are quiescent unless stimulated; in these experiments they are driven at a frequency of 0.5 stimuli/sec. The sinus venosus preparations are spontaneously active.

All of our electrodes K^+, Cl^- and KCl filled, are made from pipettes pulled from Pyrex (Corning code 7740) tubing at the same settings on the pipette puller. They have a nominal tip diameter of 0.5 μ and have a resistance in the range of 10-12 megohms when they are filled with saturated KCl. We use the same pipettes throughout because we have found a shape that is suitable for penetrating cardiac muscle cells. The fabrication of the K^+ and Cl^- electrodes has been described previously (1). The reference electrode for all measurements is a Ag-AgCl electrode with a saturated KCl salt bridge.

Intracellular Potassium Measurements

On impaling a cell with a K^+ microelectrode and a sweep speed of 200 msec/cm there is a square-step change in electrode potential. This change in potential, ΔE, is composed of two parts as shown in equation 1. The first part on the right side of the equation is the membrane potential, E_M. By membrane potential we mean the diastolic

$$(1) \qquad \Delta E = E_M + \frac{RT}{F} \log_e \left[\frac{a_K^i}{a_K^o + k_{K,\, Na}\, a_{Na}^o} \right]$$

potential, i.e. the steady or resting potential between the action potentials. In ventricle and atrium this is a steady potential that very abrubtly gives way to an action potential and then returns to the resting potential. In sinus venosus the resting potential undergoes a slow depolarization that gives way, more or less rapidly, to an action potential, depending on whether it is a primary or a secondary pacemaker cell. E_M is taken as the lowest point during the diastolic phase.

The second part on the right is the Nicolski equation (2). This describes the change in potential due to the difference in potassium activity on the two sides of the cell membrane. The denominator is the potassium activity outside of the cell (a_K^o) plus the sodium activity (a_{Na}^o) multiplied by the selectivity coefficient, $k_{K\ Na}$. We never use electrodes for which the selectivity coefficient is larger than 0.02, i.e., when the K:Na selectivity is less than 50:1. In the numerator only the potassium activity appears because we assume that the product of the selectivity coefficient and sodium activity is so small by comparison with the intracellular potassium activity that it can be neglected.

From this equation it can be seen that it is necessary to make two measurements in order to determine the intracellular potassium activity. You have to measure ΔE with the ion selective electrode and then you have to measure E_M with a conventional saturated KCl filled pipette. Ideally this would be done with a double barrel pipette, one barrel being the K^+ electrode and the other one the saturated KCl filled for measuring the membrane potential. To date we have not been able to construct a double barrel electrode that is satisfactory. The K^+ barrel usually works very well but the reference side always has a tip potential that is too large for it to be useful for intracellular recording. What we have done is to make 5-10 measurements with each of the two kinds of electrodes in every preparation, then used the average values from these two sets of measurements to calulate the intracellular potassium activity. It should be noted, however, that if you are only interested in the difference between E_M and E_K (the potassium equilibrium potential), that number can be obtained from ΔE alone. Using the definition of E_K, equation 1 can be written in the form shown as equation 2 in which the last term on the right side is known:

$$(2) \qquad \Delta E = E_M - E_K + \frac{RT}{F} \log_e \left[\frac{a_K^o}{a_K} \right]$$

TABLE I

	Sinus Venosus	Atrium	Ventricle
ΔE(mv)	+17.3 ± 0.56 (109)	-12.0 ± 0.23 (200)	-12.5± 0.20(211)
$E_M - E_K$ (mv)	+32.7 ± 0.53 (109)	+2.9+ 0.23 (200)	+2.7 ± 0.19(211)
E_M (mv)	-54.8 ± 1.63 (27)	-91.3 ± 0.43(190)	-93.8 ± 0.57(195)
E_K (mv)	-88.5 ± 1.48 (7)	-94.2 ± 0.46(20)	-96.2 ± 0.55(19)
a_K^i (mM/L)	64.4 ± 3.82 (7)	79.8 ± 1.45 (20)	86.2 ± 0.65(19)

TABLE I. Data from intracellular measurements in frog heart muscle made with potassium microelectrodes and KCl filled micropipettes. Entries are mean values plus or minus the standard error of the mean.

The numerator is the real potassium activity in the outside solution and the denominator is the apparent potassium activity as measured with the potassium electrode. Since both of these quantities are known the difference $E_M - E_K$ can be calculated from ΔE.

Table I contains the potassium data that we have obtained. Each entry in the Table is a mean plus or minus the standard error of the mean. The numbers in the brackets are the numbers of measurements. For example, under sinus venosus the number 109 in brackets after the mean value for ΔE indicates that there were 109 potassium electrode measurements and the 7 following E_K shows that those measurements were made in a total of 7 different preparations. The first row across the top is the values of ΔE in the three kinds of heart muscle. As you can see, the values of ΔE are large enough so there is no doubt when you penetrate a cell. The fact that it is not a perfect potassium electrode, i.e. the sodium in the outside solution contributes substantially to the electrode potential, actually helps you by making the ΔE value larger than it otherwise would be.

In the bottom row are the calculated values of intracellular potassium activity, it is highest in ventricle, lower in atrium and lowest in sinus venosus. The ranges of values are not shown but the activity varies by about a factor of two in the group of animals in which we made measurements. For example, in ventricle it varied from a low of 50 mM/L up to about 100 mM/L. In the second row from the bottom are the calculated values of E_K, and finally in the second row from the top the most interesting data, the difference between E_K and E_M. You can see that in all three tissues E_K is more negative than E_M. In ventricle and atrium this difference is slightly less than three milivolts and is remarkably constant despite the variation in a_K^i from one preparation to another. In sinus venosus, however, the situation is very different. Here E_K is more than 20 mv negative with respect to E_M. The differences between E_M and E_K in these tissues are about as expected, but this is, as far as we know, the first time the expectations have been confirmed by direct measurements.

Now what other information can be obtained from the data? One obvious relationship to explore is the plot of E_M as function of log a_k^i. For ventricle the results are unambiguous. A least squares linear regression plot yields a slope between 56 and 57 mv, which is not significantly different than 58 mv. The correlation coefficient is 0.91. Atrium and sinus venosus presented a rather disquieting picture in that the slopes of the straight lines are about one half Nernstian. The correlation coefficient for atrium is lower, 0.72, but for sinus venosus the correlation coefficient is again 0.91. From these data it appears that the relation is a linear one but with only one half Nernstian slopes.

We next explored the fit of the data to a more detailed relationship than that forming the basis of Nernst equation. We wrote the Goldman constant field equation (3) in the form used by Hodgkin and Katz (4) in the exponential form shown here as equation 3.

$$(3) \qquad \exp\left(-\frac{E_M F}{RT}\right) = \frac{a_K^i + \alpha\, a_{Na}^i + \beta\, a_{Cl}^o}{a_K^o + \alpha\, a_{Na}^o + \beta\, a_{Cl}^i}$$

α is the ratio of the sodium permeability to the potassium permeability and β is the ratio of the chloride permeability to the potassium permeability. If one makes the rather crude assumption that everything on the right side of the equation is constant except a_K^i, the equation is simplified to equation 3a where, when $\exp \frac{-E_M F}{RT}$ is plotted as a function of a_K^i a straight line is obtained with slope 1/B and an intercept on the ordinate A/B.

$$(3a) \qquad \exp\left[-\frac{E_M F}{RT}\right] = \frac{a_K^i + A}{B}$$

Figure 1 shows the data for the ventricle plotted in this way. The correlation coefficient is still 0.91. The values of A and B can be obtained from the exponential plots and used to calculate E_M for a range of values of a_K^i. If you then plot E_M as a function of log a_K^i, you get a curved line.

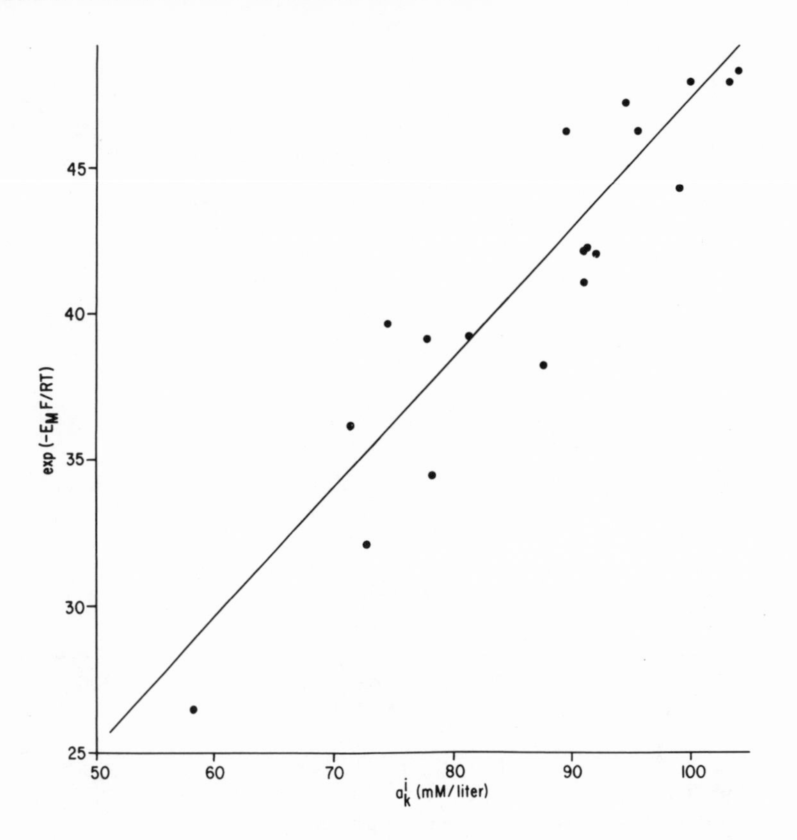

Figure 1. Plot of exp ($-E_M F/RT$) as a function of a^i_K for frog ventricle. The line is the least squares linear regression calculated from the data points.

Figure 2 shows the plot for atrium. The solid line is the least squares straight line. It has a slope of about 30 mv. The dash line is calculated from the constant field equation using the values of A and B derived from the exponential plot. Obviously the values of a^i_K do not extend over a large enough range of values to make it possible to say that one curve fits the data better than the other. It is convenient therefore to simply say that the data are fit by the constant field equation.

Figure 3 shows the plot of the same relationships for sinus venosus. The solid line is the straight line with about a 30 mv slope and the dash line is calculated from the constant field equation.

By assuming values for intracellular sodium and chloride activities one can make order of magnitude calculations for the values

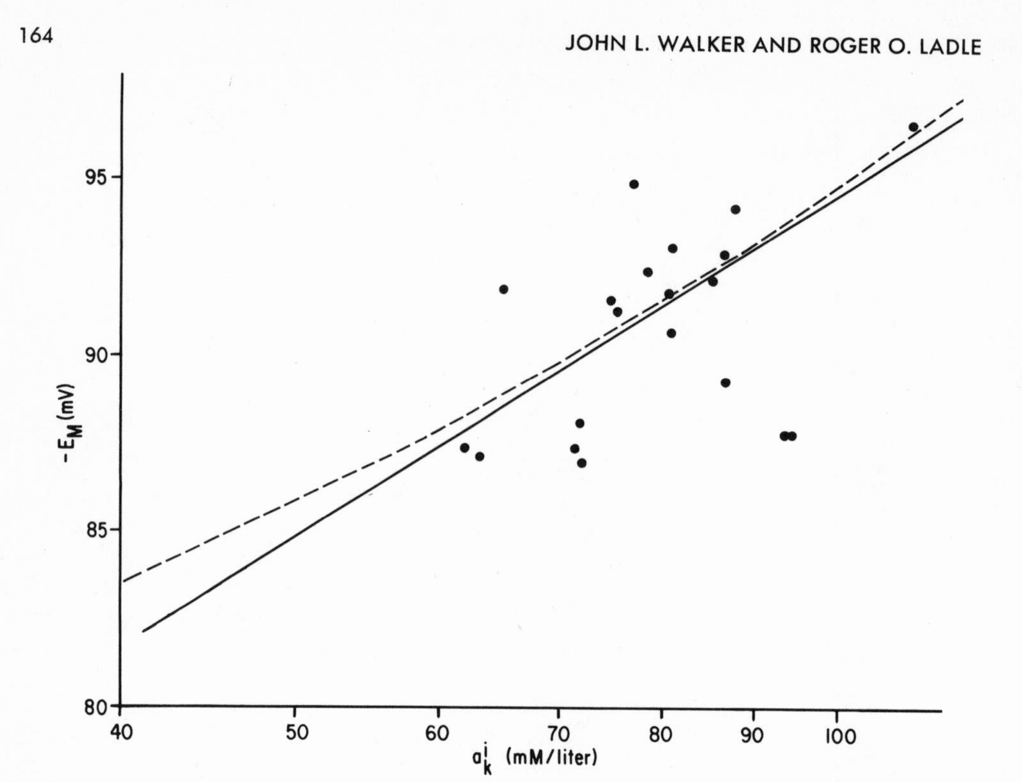

Figure 2. Plot of $-E_M$ as a function of $\log_{10}a_K^i$ for frog atrium. The solid line is the least squares linear regression. The dashed line is the constant field equation fit to the data.

of α and β. The results show that α and β are relatively insensitive for the values used for these activities, at least within what seems to be the most likely range. The values of $\alpha(P_{Na}/P_K)$ in Table II are about the same for atrium and ventricle, while it is an order of magnitude larger in sinus venosus . This finding is consistent with the observation that E_M is close to E_K in atrium and ventricle, but E_M is very different from E_K in sinus venosus. The data also support the physiological concepts that have existed for a long time, that the sinus venosus has a higher sodium permeability relative to potassium than do atrium and ventricle, and that the sodium permeability therefore makes a significant contribution to E_M in the sinus venosus.

Looking at β one can see that it is an order of magnitude higher in the sinus venosus and atrium than in the ventricle. This may be the explanation for the difference in slopes of the plots of E_M vs log a_K^i. Ventricle, which has the lowest P_{Cl}/P_K ratio, shows a Nernst slope, while sinus venosus and atrium have approximately one half Nernstian slopes because the chloride permeability is

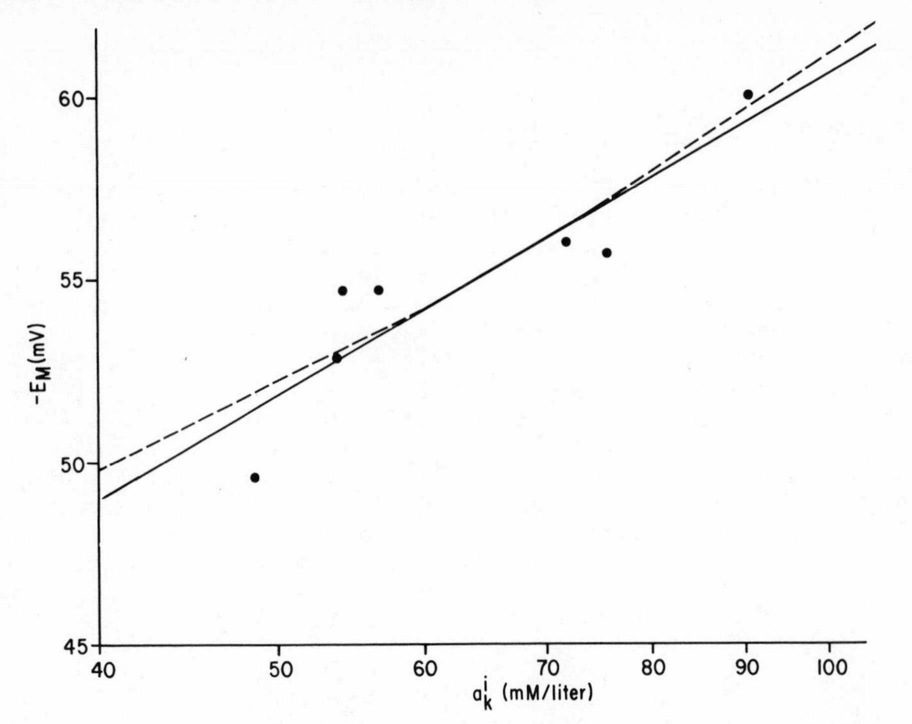

Figure 3. Plot of $-E_M$ as a function of $\log_{10}a_K^i$ for frog sinus venosus. The solid line is the least squares linear regression. The dashed line is the constant field equation fit to the data.

TABLE II

	Sinus Venosus	Atrium	Ventricle
α	$1.8 \times 10^{-2} - 1.3 \times 10^{-1}$	1×10^{-3}	$4 \times 10^{-4} - 2 \times 10^{-3}$
β	$7.0 \times 10^{-1} - 7.6 \times 10^{-1}$	7.6×10^{-1}	7.4×10^{-2}

TABLE II. Values of the permeability ratios α and β estimated from the constant field equation fitted to the intracellular potassium data.

sufficiently high to affect the slope of the potassium response.

Intracellular chloride activities

Chloride activity measurements are done in the same manner as the potassium measurements, except that the Nernst equation was used instead of the Nicolski equation to describe the potential difference due to the difference of chloride activities between the inside and outside of the cell. We know that there are no interfering ions in the solution on the outside. We assume that there are no interfering anions inside the cell. This assumption is necessary because we simply do not have any information about what ions are present and what their activities are.

The data in the top row in Table III clearly show that (as with the potassium electrode) the value of ΔE is large enough so that there is no problem telling when you are inside of the cell with the chloride electrode. a^i_{Cl} in all three tissues is much higher

TABLE III

	Sinus Venosus	Atrium	Ventricle
ΔE (mV)	-32.4±1.27 (52)	-48.7±1.74(116)	-50.4±53(120)
E_M (mV)	-54.8±2.05 (66)	-89.8±1.82(105)	-90.4±0.61(117)
a^i_{Cl} (mM/L)	38.0±3.79 (10)	17.2±2.0 (11)	17.6±0.57(12)
E_{Cl} (mV)	-23.4±2.71(10)	-42.5±2.8(11)	-41.9±0.81(12)
E_M-E_{Cl} (mV)	-31.4	-47.3	-48.5

TABLE III. Data from intracellular measurements in frog heart muscle made with chloride microelectrodes and KCl filled micropipettes. Entries are mean values plus or minus the standard error of the mean.

than we expected on the basis of a Donnan equilibrium distribution at E_M. If chloride were passively distributed at E_M, the activity should be less than 2mM. However, a^i_{Cl} is approximately ten times higher than predicted on that basis. Of course, since it is so high, the calculated E_{Cl} is much more positive than E_M. In the bottom row the difference between E_M and E_{Cl} is shown.

When E_M is plotted as a function of log a^i_{Cl} and the points are fitted with a straight line, the slope is very much lower than Nernstian. For ventricle, atrium and sinus venosus the slopes are 17.5 mv, 20.7 mv and 34 mv, respectively. As with the potassium data one can use an exponential plot to derive parameters to fit the data with a constant field equation. Figures 4 through 6 show this plot for venticle, atrium and sinus venosus respectively. In each figure the solid line is the least squares straight line, and the dashed line is calculated from the constant field equation using the parameters derived from the exponential plot. In every case the dashed line fits the data just as well as the straight line.

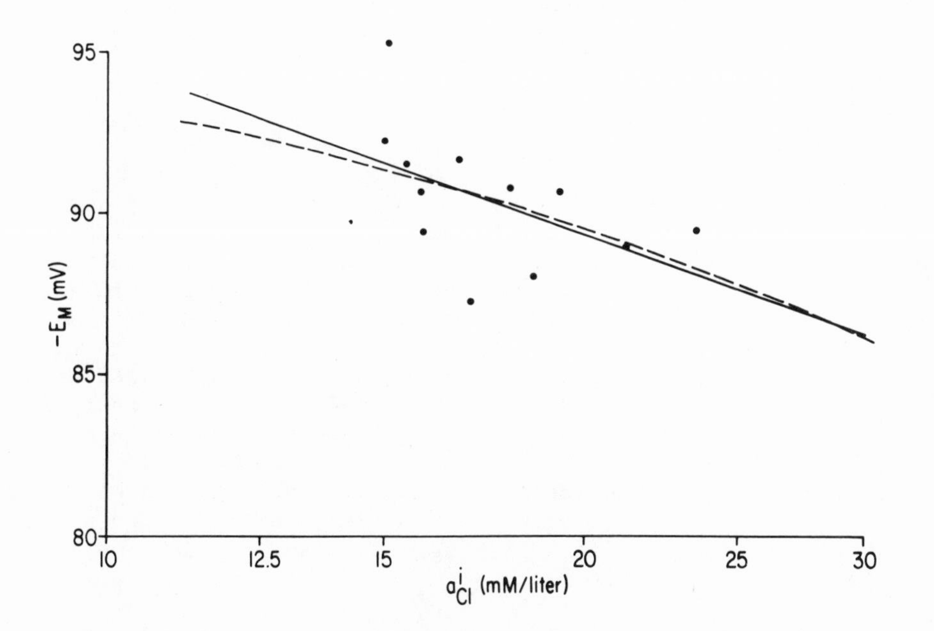

Figure 4. Plot $-E_M$ as a function of $\log_{10}a^i_{Cl}$ for frog ventricle. The solid line is the least squares linear regression. The dashed line is the constant field equation fit to the data.

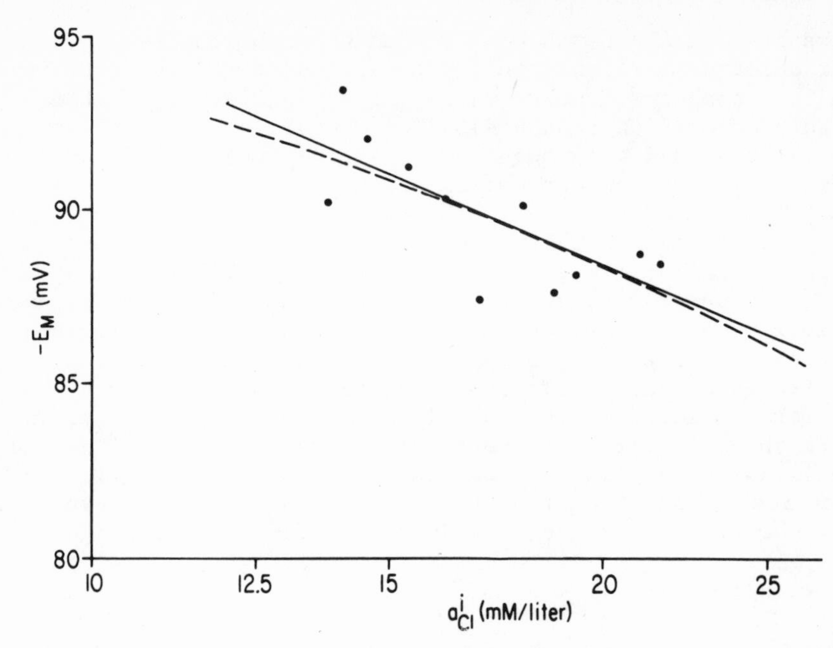

Figure 5. Plot $-E_M$ as a function of $\log_{10} a^i_{Cl}$ for frog atrium. The solid line is the least squares linear regression. The dashed line is the constant field equation fit to the data.

The problem with chloride measurements is that the chloride electrode is not very selective. It sees organic anions. There is therefore a real question as to whether what we are measuring is chloride or chloride plus something else.

So far the only successful control we have done is to take five frog hearts, wash them with isotonic sucrose to get rid of most of the extracellular chloride, and then make a pooled homogenate of them. Chloride activity in the homogenate was then measured with a liquid ion exchanger chloride microelectrode and with a Ag-AgCl electrode. The measurements with the two electrodes agree to within 1 mM. In view of this, we feel that we are measuring the true intracellular chloride activity.

We are now trying to make intracellular measurements with the recessed tip Ag-AgCl microelectrodes described by Thomas (5). To date we have not had a complete experiment, but the few measurements that we have obtained with the Ag-AgCl microelectrode agree

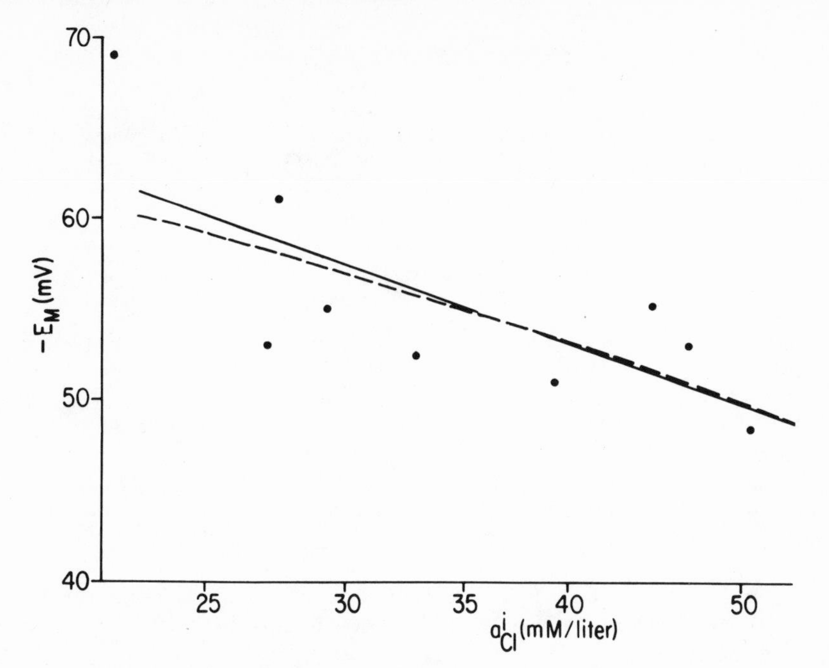

Figure 6. Plot of $-E_M$ as a function of $\log_{10} a^i_{Cl}$ for frog sinus venosus. The solid line is the least squares linear regression. The dashed line is the constant field equation fit to the data.

quite well with the liquid ion exchanger microelectrode measurements.

If these measurements do turn out to be accurate, then the next question is: What is the mechanism of the chloride distribution? Is it actively transported into the cells, or is it, as has been suggested, passively distributed at some potential that is obtained by taking a time average of the membrane potential during the cardiac cycle? We have some evidence from experiments with atrium and ventricle that chloride may not be passively distributed. In these experiments we measured the chloride activity in a beating preparation, then turned the stimulus off and waited for at least two hours before making another set of measurements of chloride activity in the non-beating preparation. These experiments showed that the difference between E_M and E_{Cl} does not change significantly after the preparations have been quiescent for as long as six hours. This investigation is continuing at the present time.

Acknowledgment

This work was supported by a grant-in-aid from the American Heart Association and by N.I.H. grant NS 07938.

References

1. Walker, J.L. "Ion Specific Liquid Ion Exchanger Microelectrodes" Anal. Chem. 43: 89A (1971).

2. Nicolsky, B.P., "Theory of the Glass Electrode I", Acta Physicchem. USSR, 7:597 (1937).

3. Goldman, D.E., "Potential, Impedance, and Rectification in Membranes., J.Gen. Physiol. 27:37-60, 1943.

4. Hodgkin, A.L. and Katz, B., "The Effect of Sodium Ions on the Electrical Activity of the Giant Axon of the Squid"., J. Physiol. 108: 37 (1949).

5. Neild, T.O. and Thomas, R.C.," New Design for a Chloride-Sensitive Microelectrode", J. Physiol., 231:7 (1973).

DISCUSSION

Question (Krnjevic)

1. Is gCl^- in your preparation pH sensitive as in other muscle?
2. Have you tried varying Na^+ concentrations to see whether it is gNa that is high in the sinus venosus and not gCa?

Answers (Walker)

1. I have no information about that.
2. No.

Question: (Durst)

Is there some biological or other reason for using the Corning ion exchanger for potassium rather than valinomycin which would be completely insensitive to the sodium interference?

Answer: (Walker)

Not that I am aware of - The Corning exchanger was the first one available to me and it has been so satisfactory that I have not bothered to try valinomycin. The selectivity of the

Corning exchanger in the microelectrode is at least 50 in favor of K^+ over Na^+ and that is more than enough for my purposes.

Question: (Whalen)
Have you tried superfusing the sinus venosus preparation with a solution containing acetylcholine at different concentrations as a kind of voltage-clamp device? I think it could be done but whether it would help solve the chloride story I have no idea.

Answer: (Walker)
We haven't tried it because we are not able to get into and hold cells for more than one or two minutes, so, unless the preparation can be perfused for several hours with acetylcholine, it is not an experiment we can do. We have not looked into the feasibility of using acetylcholine on such a long time basis.

Question: (Hollander)
Experiments using Ouabain to poison active transport of ions, coupled or otherwise, may help to elucidate the activity and permeability of Cl^- ions.

Answer: (Walker)
It is a possibility but until we are able to do the proper control experiment with the recessed tip Ag-AgCl electrode of the Thomas type to verify our Cl^- measurements made with the Cl^- exchanger electrode we are not going to do anything else. If we can verify our results we will of course go to such things as drugs and cooling to see if we can push the a^i_{Cl} around.

Question: (Brown)
Can you give a comparative statement of the efficacy of the various techniques for "siliconizing" micropipette tips?

Answer: (Walker)
Almost any of chloro- or methoxysilanes are suitable for treating pipettes to be used for K^+ - electrodes if 50:1 K^+ over Na^+ selectivity is sufficient. For higher selectivities with the K^+ electrode and for the Cl^- electrode the Dow-Corning 200 fluids are best. The problem is that the concentration is very critical. Too little and the pipette won't hold the oil, too much and the pipette will be plugged. A compromise has been to use siliclad. The concentration is important but not as critical as the DC-200 fluids.

VI. Clinical Applications

MUSCLE pH, PO_2, PCO_2 MONITORING:

A REVIEW OF LABORATORY AND CLINICAL EVALUATIONS

Robert M. Filler, M.D., John B. Das, M.D., Ph.D.

Department of Surgery, Children's Hospital Medical Center, and Harvard Medical School
300 Longwood Avenue, Boston, Massachusetts 02115

INTRODUCTION

The vital physiologic data necessary for the care of the critically ill are currently obtained from clinical observation, pulse, arterial and central venous pressures, electrocardiogram and blood gas determinations. In the infant, these parameters are often inaccurate (clinical observation), difficult to obtain because of the infant's size (blood pressure), potentially hazardous (in-dwelling arterial line), and fail to provide minute-to-minute information (blood gas determination). In addition, these parameters may not necessarily reflect the physiologic state of the peripheral tissues. In our search for better and safer methods to assess vital functions in the very young, we have been evaluating the feasibility of continuous tissue monitoring of pH and more recently PCO_2 and PO_2.

MUSCLE pH MONITORING

The rationale for the use of surface or extracellular muscle pH measurements as a guide to important physiological variables is straight-forward. The acidity of the surface fluid over a muscle corresponds to the acidity of the blood entering the muscle under most circumstances. However, if oxygen supply to a muscle is inadequate, anaerobic glycolysis results in lactic acid production. The acid diffuses into the extracellular fluid surrounding the muscle and lowers surface pH. Therefore, muscle surface acidosis would be expected in states of arterial acidosis and when oxygen supply to muscle is reduced i.e. as a result of poor vascular perfusion or a decrease in oxygen content of the perfusing blood. The reports of Lemieux, Smith, Couch (1,2) first showed that the pH on the surface of skeletal muscle was indeed a reliable index of both arterial pH and tissue perfusion in dogs subjected to acidosis, alkalosis and hemorrhage. Because these findings suggested that continuous measurements of muscle pH could provide just the kind of physiologic information that is needed in the very ill patient, and because relatively simple manipulations were necessary to obtain such data, the method seemed particularly attractive for use in critically ill infants.

About 3 years ago (with the assistance of Corning Glass, Medfield, Mass.) a combination glass pH electrode was designed which could be placed on the surface of an extremity muscle in the smallest infant. This probe has a glass membrane for sensing and a Ag-AgCl internal reference with a ceramic junction. This probe is 7 cm. in length and 0.7 cm. in diameter. The sensing tip is at the distal end of a glass shaft and the reference junction is 1 cm. proximal to it. The electrode can be sterilized with ethylene oxide. Its 10-foot cable allows convenient placement of the pH meter to which it is attached for continuous reading.

Animal Studies

Initial animal studies quickly confirmed the efficacy of equipment and the physiologic information obtained encouraged us to proceed with clinical trials (3). Typical muscle pH responses in the experimental animal are shown in Figures 1, 2 and 3. In these studies, the oxygen supply to the muscle being monitored was decreased by vascular occlusion, hemorrhagic shock or arterial hypoxia.

Clinical Experience (4)

To date, more than 100 sick infants and children have had a pH electrode inserted for continuous monitoring. The average duration of monitoring has been over 50 hours, and a single probe has been left in situ for as long as 21 days. Total monitoring experience has exceeded 5000 hours.

The technique of inserting the probe in an infant is shown in Figure 4. Except for one patient, the vastus lateralis muscle was used for monitoring in all. To avoid electrical hazards which might be associated with the use of line powered pH meters, a portable battery operated model is always used when patients are being monitored. No wound infections have been noted at the probe site and after withdrawal of the probe, wound healing has been satisfactory in all. The drift in calibration of the probe which occurred during use rarely exceeded 0.05 pH units. Temperature correction factors recommended for pH readings at temperatures other than 37° C. are small enough to be clinically insignificant and can be ignored during the routine use of the probe, even in a hypothermic child.

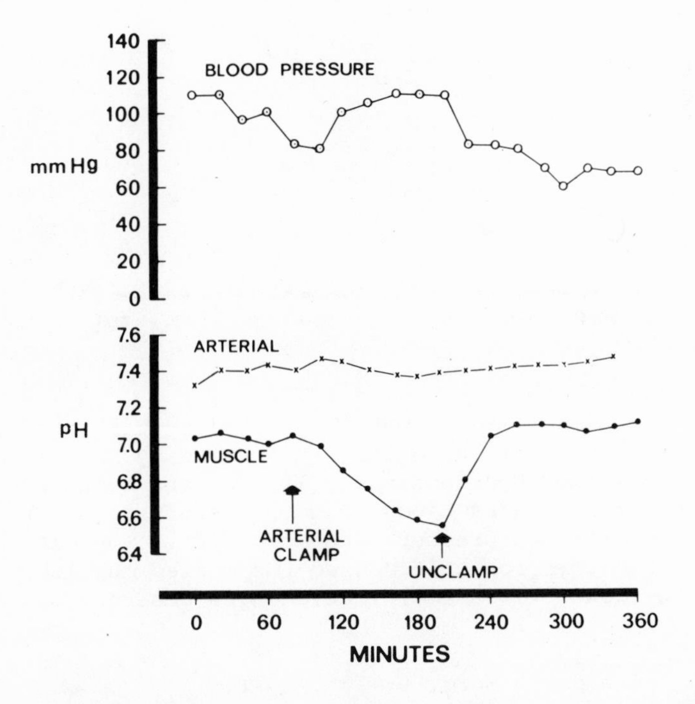

Figure 1 - Effect of limb ischemia on muscle pH. The muscle pH in this dog's limb fell rapidly from 7.10 to 6.55 when its arterial blood flow was clamped for 2 hours. pH rose promptly with restoration of circulation. Central arterial pH was unaffected by the acidosis on the muscle surface of the poorly perfused limb. Mean arterial pressure was stable.

Reprinted from J. of Pediat. Surg. 6: 535, 1971.

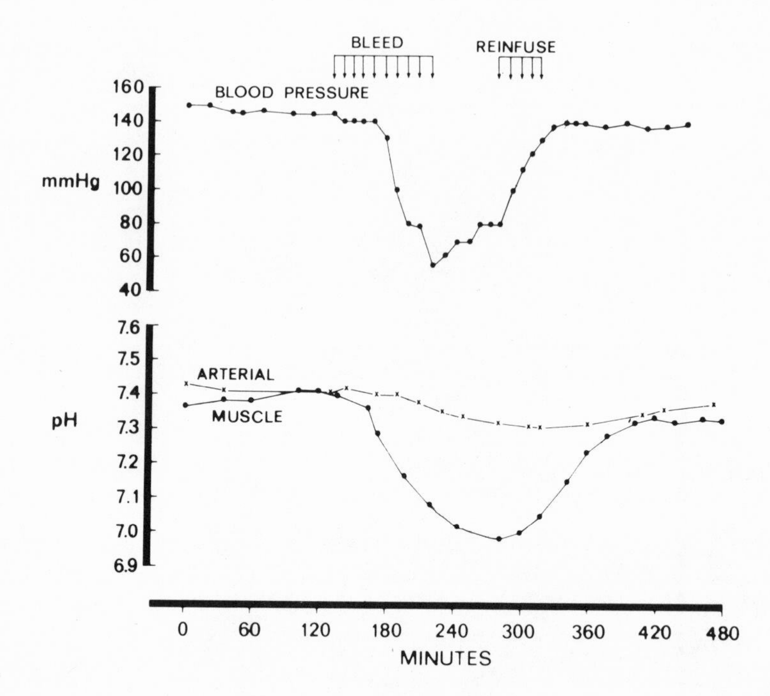

Figure 2 - Effect of hemorrhage on muscle pH. Blood was withdrawn from this 12 kg. dog in 50 cc. aliquots represented by each arrow. Muscle pH decline preceded hypotension by 15 min. and began after the removal of 15% of estimated blood volume. Reinfusion of shed blood restored mean arterial pressure and muscle pH. The extent of peripheral vasoconstriction that caused the marked muscle acidosis was not evident from the minimal changes observed in arterial pH.

Reprinted from J. of Pediat. Surg. 6: 535, 1971.

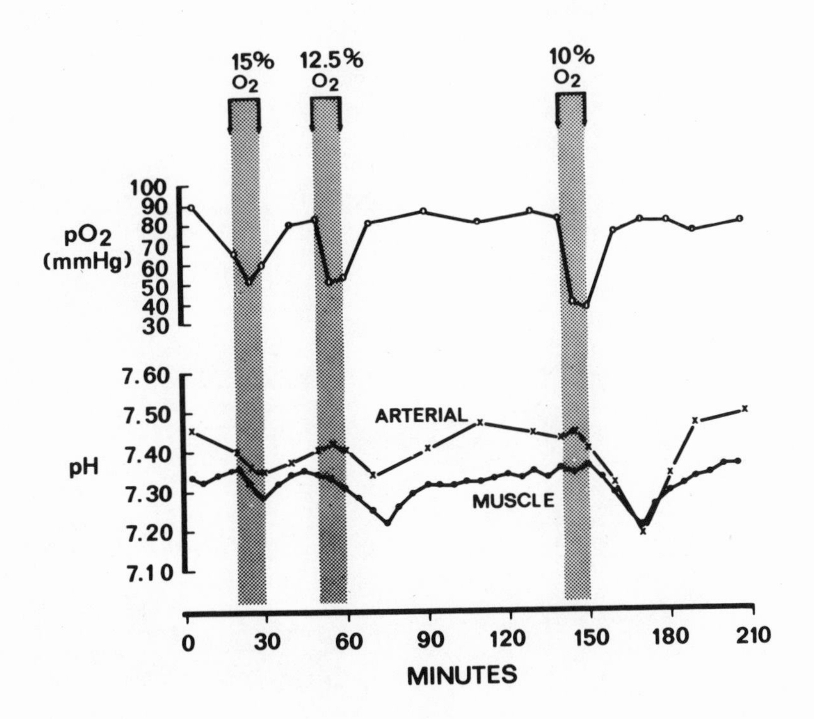

Figure 3 - Effect of hypoxia on muscle pH. In this dog, the percentage of inspired oxygen was reduced from 20% (room air) to 15, 12.5, and 10% during the 5-minute intervals illustrated. Arterial PO_2 and pH responded promptly to changes in inspired oxygen concentrations. The muscle pH responses were similar in magnitude with a time lag of about 5 minutes.

Reprinted from J. of Pediat. Surg. 6: 535, 1971.

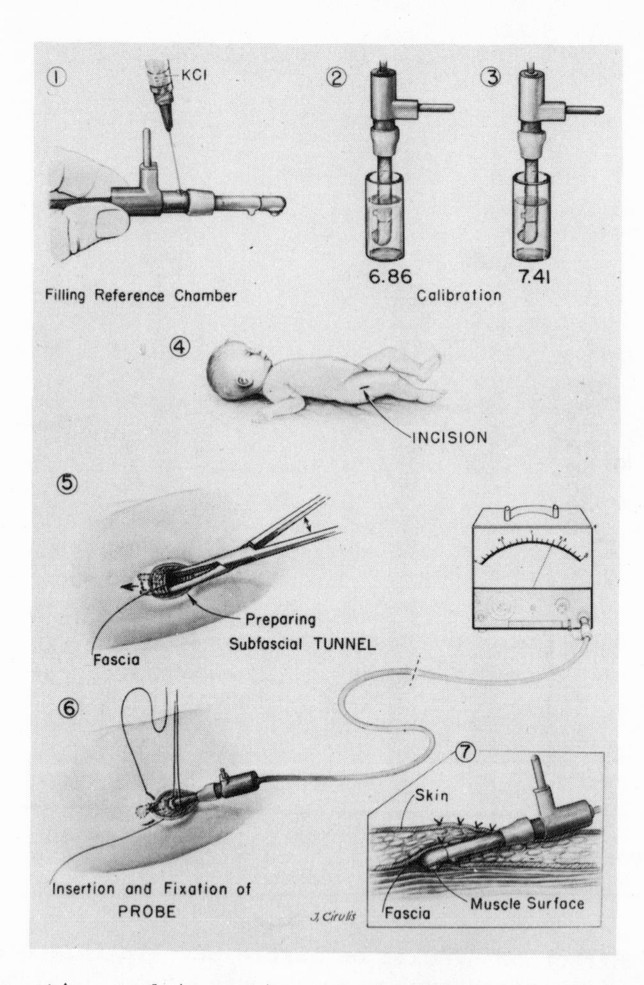

Figure 4 - Preparation and insertion of pH probe. 1) The residual 4M KCL solution in the reference chamber is aspirated and replaced with fresh solution. 2 and 3) The sterile probe is calibrated in low (6.86) and high (7.41) standard pH buffers. 4) A 2 cm. incision is made down to the fascia lata in which a 1 cm. incision is made. 5) A subfascial tunnel which will accept the distal 2 cm. of the probe is made with blunt dissection. 6) The probe, the cable of which is attached to a pH meter, is stabilized by closing the fascia and skin over the probe and taping the exposed portion of the probe to the skin. 7) Final position of the probe on the muscle surface. The right angle extension on the exposed plastic cap is useful for securing the probe to the limb and orienting the sensing tip.

Reprinted from Surgery 72: 23, 1972.

Our experience indicates that the pH on the surface of skeletal muscle normally ranges from 7.35 to 7.45. We have learned that values outside of this range are best interpreted by comparing them to blood pH. Three categories of abnormalities have been defined: 1) Low muscle pH - normal blood pH, 2) Low muscle pH - low blood pH, 3) High muscle pH.

Low muscle pH with normal blood pH, the most common abnormality seen, occurs in patients with clinical problems which cause a decrease in tissue perfusion such as bleeding, extracellular fluid deficit, decreases in cardiac output, and any condition which causes peripheral vasoconstriction and/or vasculitis. In children with decreased tissue perfusion, muscle pH as low as 6.60 has been observed. The lowest values have occurred during massive hemorrhage. Often muscle pH decline in hemorrhage, a result of reflex peripheral vasoconstriction, occurs before a fall in central blood pressure. The degree of muscle acidosis is related to the rate of active bleeding and the ability of transfusions to keep pace with losses. As a result, muscle pH is an excellent index of adequacy of transfusion. Similarly in other clinical situations associated with poor muscle perfusion such as a child with large gastrointestinal fluid losses or a hypothermic infant, the effectiveness of therapy can be judged by the rise in muscle pH. In these clinical settings, lactic acid "wash-out effect" has been noted in the children in whom blood lactate was measured. Blood lactate is generally normal or only slightly elevated during periods of severe muscle acidosis due to poor perfusion. When adequate circulation is restored, muscle pH rises and a marked increase in blood lactate is noted as the lactic acid which has accumulated in the tissues during shock is washed into the blood stream. The use of muscle pH monitoring for diagnosis and management of a child with persistent post-operative bleeding is shown in Figure 5.

Low muscle pH associated with low blood pH occurs in patients with either respiratory or metabolic acidosis. Blood PCO_2 readily differentiates respiratory from metabolic causes. Muscle pH values in these conditions have been as low as 6.60. We have now seen 6 infants who unexpectedly developed low muscle pH during an "uneventful" course under general anesthesia. Muscle acidosis suggested the need for blood gas analysis which proved that each infant had a respiratory acidosis due to inadequate ventilation. In each case, a change in anesthetic technique improved the ventilation and muscle pH (and blood gases) rapidly returned to the normal range. In Figure 6, the relation of muscle pH and blood pH, and PCO_2 in an infant with respiratory acidosis is shown. This example also shows that the effectiveness of therapy can be judged on a minute-to-minute basis by observing the continuous muscle pH readings.

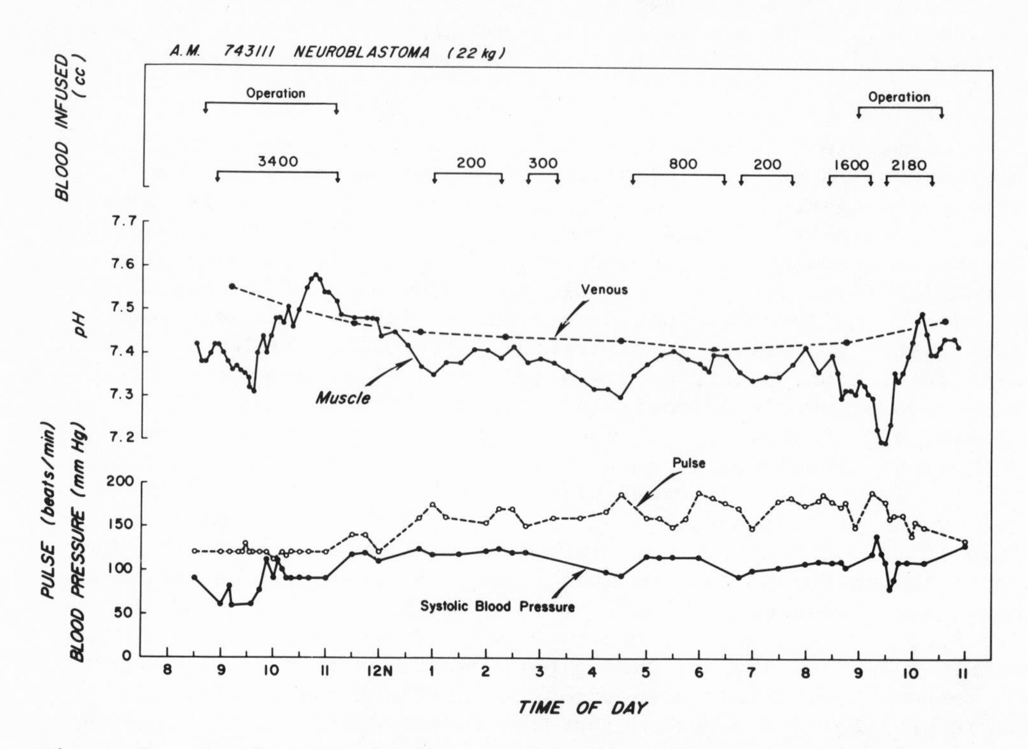

Figure 5 - Muscle pH recording during and following excision of large right adrenal neuroblastoma which extended into the liver in a boy weighing 22 kg. (Patient H.M.). At the first operation between 8:30 and 11:00 a.m., extensive operative blood loss occurred. With moderate hypotension, the initial muscle pH value (7.40) fell to 7.30, 0.2 lower than venous pH (elevated because of hyperventilation), indicating a moderate decrease in tissue perfusion. Muscle pH rose to blood level after satisfactory blood replacement (3,400 cc. - 1.8 blood volumes). During the afternoon, muscle pH slowly fell (in the absence of hypotension or change in blood pH) but responded to transfusions at 1:00 and 4:00 p.m. By 8:00 p.m., the abdomen was distended, and exploration for bleeding was undertaken. Upon the opening of the abdomen, the muscle pH fell dramatically in the absence of any acute bleeding. A blood clot was removed, several small vessels were ligated and more blood was infused. Muscle pH responded and rose to blood levels. The child made an uneventful recovery.
Reprinted from Surgery 72: 23, 1972.

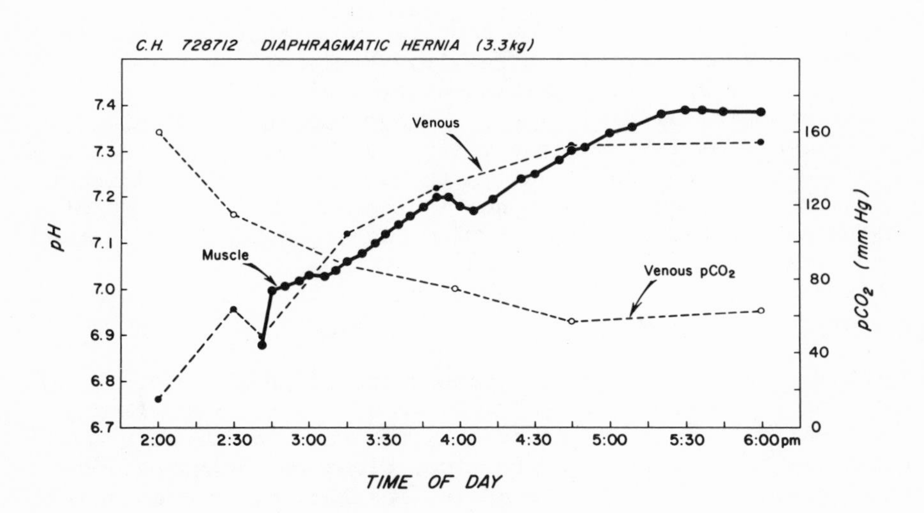

Figure 6 - Continuous muscle pH data recorded every 5 minutes during treatment of respiratory acidosis in a newborn infant weighing 3.3 kg. (Patient C. H.), with left Bochdalek diaphragmatic hernia. Muscle pH monitoring was started at 2:30 p.m., 2 hours following completion of the operative repair when blood gas data confirmed that the patient had a severe respiratory acidosis (blood pH, 6.78 and PCO_2, 160 mm. Hg.). The baby was treated with a ventilator (60 percent oxygen) and muscle pH responded appropriately as confirmed by sampling of central venous blood every 60 to 90 minutes.

Reprinted from Surgery 72: 23, 1972.

High muscle pH (above 7.45) is seen in children with a respiratory or metabolic alkalosis. As in acidotic states, respiratory and metabolic causes can be distinguished by the blood PCO_2. Most of the children who have developed muscle alkalosis were alkalotic because of hyperventilation while on a mechanical ventilator or because of a metabolic alkalosis caused by the transfusion of large quantities of citrated blood. Muscle pH which is often higher than blood pH in these children has been observed as high as 7.70.

A major feature of muscle pH monitoring is that information is displayed continuously. As a result, muscle pH monitoring serves as an important early warning system, and in addition to some of the examples already given, we have had numerous others in which a muscle pH fall was the first sign of patient deterioration. In addition, we found that in the absence of a significant change in muscle pH, no major change in the hemodynamic or respiratory status of a patient has occurred. As already noted, minute-to-minute data are also extremely useful to evaluate and modify therapy especially when therapy can produce significant changes rapidly.

Although blood sampling for measurement of pH and PCO_2 is essential for the accurate interpretation of abnormal muscle pH, when muscle pH monitoring is in effect, the need for frequent blood sampling is reduced. In practice, blood pH and gases are measured when an abnormal muscle pH is first noted, when an unexpected fall in muscle pH greater than 0.05 occurs, or when therapy does not produce the expected improvement in muscle pH. When the muscle pH value is stable in the normal range, measurement of blood pH is usually not necessary.

Early in our experience, frequent blood pH and gas determinations were made and blood samples were almost always obtained from arterial lines. Experience has shown that central venous blood pH and PCO_2 are satisfactory for the interpretation of muscle pH. The use of central venous blood samples eliminates the need for indwelling arterial lines which may prove hazardous, especially in the small infant.

One of the disadvantages of the present method of pH monitoring is that an incision is necessary to implant the rather large probe that is currently available. We are now working to develop a pH electrode which can be inserted percutaneously. Such an improvement in hardware will not require surgical expertise for implantation and will make muscle pH monitoring available to many more patients.

MUSCLE PCO_2 AND PO_2 MONITORING

Recent work has shown that mass spectrometry can be applied to continuous in-vivo analysis of blood and tissue gas tensions (5). For tissue gas tensions, the technique requires the implantation of a teflon diffusion membrane which is attached to a hollow stainless steel tube for gas sampling and measurement. The available membrane can be passed into the tissues percutaneously through a needle. Because of the simple manipulations necessary to insert the membrane into the skeletal muscle, this method would seem technically well suited for patients of all ages and sizes.

To evaluate the clinical importance of tissue PO_2 and PCO_2, and to become familiar with this technique of monitoring, a series of animal studies were undertaken (6). In the laboratory, the teflon membrane was inserted into the rectus femoris or adductor muscle of the thigh of an anesthetized dog. In these animals, arterial blood gases were measured intermittently in vitro by standard techniques and muscle pH was measured continuously by methods already described. The magnitude and rate of change of muscle pH, PCO_2 and PO_2 and arterial pH, PCO_2 and PO_2 were determined in animals subjected to hemorrhage, hypoxia, hypo and hypercarbia, and metabolic acidosis. Data from a typical experiment in which shock was produced in 6 dogs by the withdrawal of 40-50% of the estimated blood volume and then treated by the re-infusion of shed blood are shown in Table I. Blood lactate was also measured in this study.

These data show that muscle PO_2 falls, PCO_2 rises and muscle pH falls due to poor muscle perfusion caused by hemorrhagic shock. The table does not show the additional observation that the fall in muscle PO_2 precedes the fall in pH and rise in PCO_2 and occurs when 10% of the animal's blood volume is withdrawn. Changes in PCO_2 and pH occur only after 20% of the animal's blood volume is removed. Arterial acidosis occurs late, and as expected, arterial PO_2 and PCO_2 change very little early in the course of bleeding. The evolutionary changes in the tissue PO_2, PCO_2, pH and blood lactate suggests the following sequence of events. Peripheral vasoconstriction occurs in response to hypovolemia. Under-perfusion of muscle leads to greater extraction of oxygen from the blood which does arrive at the periphery. (Muscle PO_2 falls.) Oxygen supply fails to keep up with the demands of the muscle and anaerobic glycolysis with production of lactic acid occurs. (Muscle pH falls, PCO_2 rises.) As circulation to the tissue is restored by transfusion, lactic acid washout is noted.

TABLE I

Average Muscle and Blood pH, PO_2, PCO_2 and Blood Lactate in 6 Dogs with Hemorrhagic Shock and Recovery

	Before Bleed	After Bleed	After Reinfusion
Mean Arterial Blood Pressure (mmHg)	112	33	94
Muscle			
pH	7.44	7.11	7.23
PO_2 (mmHg)	34	9	30
PCO_2 (mmHg)	35	50	45
Arterial			
pH	7.41	7.23	7.16
PO_2 (mmHg)	102	87	91
PCO_2 (mmHg)	29	29	36
Blood Lactate (mg%)	13	35	43

Experiments in which hypoxemia is produced, a fall in muscle PO_2 and pH parallels the fall in arterial PO_2. Likewise, the production of metabolic acidosis by the infusion of HCl causes a corresponding fall in arterial pH and muscle pH and an elevation of tissue PCO_2, but no change in muscle or blood PO_2.

These relatively simple studies have convinced us that tissue gas monitoring is practical and is capable of providing additional significant physiologic data important in the management of critically ill patients. Early clinical trials are thus far encouraging.

SUMMARY

Practical and safe methods for obtaining continuous measurement of muscle pH by an implantable probe and muscle PO_2 by mass spectrometry are currently available.

Animal studies utilizing these methods and extensive clinical trials in infants and children utilizing muscle pH measurements indicate that tissue monitoring can be extremely useful for evaluation and treatment of sick patients. The minute-to-minute readings are a continuous indication of the function of the cardiovascular and respiratory systems. As indicators of tissue perfusion, these data provide information which is not available by conventional monitoring.

BIBLIOGRAPHY

1. Lemieux, M.D., Smith, R.N. and Couch, N.P.: Surface pH and redox potential of skeletal muscle in graded hemorrhage. Surgery, 65: 457, 1969.

2. Smith, R.N., Lemieux, M.D. and Couch, N.P.: Effects of acidosis and alkalosis on surface skeletal muscle hydrogen ion activity. Surg. Gynec. Obstet., 128: 533, 1969.

3. Filler, R.M., Das, J.B., Haase, G.M., and Donahoe, P.K.: Muscle surface pH as a monitor of tissue perfusion and acid-base status. J. of Pediat. Surg., 6: 535, 1971.

4. Filler, R.M., Das, J.B., Espinosa, H.A.: Clinical experience with continuous muscle pH monitoring as an index of tissue perfusion and oxygenation and acid-base status. Surgery, 72: 23, 1972.

5. Brantigan, J.W., Gott, V.L., and Martz, M.N.: A teflon membrane for measurement of blood and intramyocardial gas tensions by mass spectroscopy. J. of Applied Physiology, 32: 276, 1972.

6. Schwartz, A., Martin del Campo, N., Hoffman, P., Crocker, D., Das, J.B., and Filler, R.M.: Tissue PO_2, PCO_2 and pH during hemorrhage, hypoxia, hypercarbia and acid infusion. (In preparation).

DISCUSSION

(Comments) (Myers)

Over the last number of years we in the Laboratory of Perinatal Physiology of NINDS, have been studying the brain pathological consequences of prenatal asphyxia using a rhesus monkey model. With severe asphyxia, arterial blood pH depressions down to 6.6-6.9 are quite common and, in fact, seem to be necessary before brain damage will occur. There has, generally speaking, existed a considerable skepticism with regards to these studies because of these severe pH deviations since most physicians have viewed such changes as incompatible with life and certainly outside the range possible in clinically meaningful situations. However, I think now the pendulum is swinging the other way and our view that such severe pH perturbations are still compatible with survival (though often not intact survival) is being vindicated. The lowest blood pH in the monkey newborn observed for 1 hour after which we have demonstrated survival was 6.57 (1).

The realization that such severe pH changes as these not only are compatible with survival but, in fact, represent the clinically significant range which relates to brain injury should hopefully stimulate instrument producing companies to develop measuring devices which operate satisfactorily in this range (2). Most currently available measuring devices fail to cover this range with accuracy.

1. Myers, R.E.: Brain damage produced by umbilical cord compression in rhesus monkey. Biol. Neonate (In Press).

2. Myers, R.E.; Threshold values of oxygen deficiency required to produce cardiovascular and brain pathological changes in fetal rhesus monkeys. In IIIrd International Conference on Oxygen Transport to Tissue. Plenum Publ. Co. H.I. Bicher and D.F. Bruley, eds. Advances Exp. Med. Biol. 37B: 1047 - 1053, 1973.

PROGRESS TOWARD A GLUCOSE SENSOR FOR THE ARTIFICIAL PANCREAS

Samuel P. Bessman and Robert D. Schultz

Department of Pharmacology
University of Southern California School of Medicine
Los Angeles, California 90033

Introduction

The artificial pancreas, actually an artificial beta cell, is a completely implantable device about the size of a cardiac pacemaker that will deliver insulin to the circulatory system, as needed, in response to increases in the glucose level of the blood or other body fluid. The device will contain a glucose sensor, a computer-amplifier system, an insulin pump, a power supply, and an insulin reservoir refillable from the outside by injection at intervals of 3 months or more. The power supply could consist of a nickel-cadmium battery and a recharging circuit activated by electromagnetic induction through the skin. When perfected, the artificial beta cell may prevent the physical deterioration of the diabetic from the stress associated with the imbalance between his actual insulin requirement at any given time and the amount of insulin in usable form in his body at that time. Much of the technology for the artificial pancreas is already available except for a reliable glucose sensor.

Enzyme Sensors

The first published report of an electrode system for continuous monitoring of glucose in biological fluids seems to be that of Clark and Lyons [1], which utilizes a layer of concentrated glucose oxidase solution trapped between two Cuprophane dialysis membranes, one of which is in contact with a pH electrode. In this system, glucose from the ambient fluid diffuses through the outer membrane and is converted to gluconic acid which then diffuses both toward the pH sensitive glass and back into the donor solution. The resultant drop in pH is a complex function of glucose concentration, the buffering capacity and flow rate of the ambient fluid, temperature and the pH. Such a system is capable of giving a response of 1 pH unit per 10 mg glucose/100 ml in a solution flowing at the rate of 3 ml per minute. However, the system has an inherently slow response related to the time required to set up steady state diffusion gradients in the dialysis membrane-enzyme solution/dialysis membrane sandwich. If, in order to speed up response, this sandwich is made very thin, the rate of diffusion of ambient buffer into the enzyme layer is also increased, resulting in a smaller pH change and a corresponding loss of sensitivity toward glucose.

A somewhat faster system suggested by Clark and Lyons [1] and later described by Clark and Sachs [2] utilizes a glucose oxidase solution trapped between a dialysis membrane and the plastic membrane of a Clark-type [3] oxygen electrode. Glucose and oxygen diffusing through the dialysis membrane react in the enzyme solution to yield gluconic acid and hydrogen peroxide. The resultant decrease of oxygen concentration in the enzyme solution is a measure of glucose concentration. The response time of this system depends on the diffusion times for glucose and oxygen from the ambient fluid through the dialysis membrane and through the enzyme layer and also the time for oxygen to diffuse through the plastic membrane and the electrolyte layer adjacent to the polarized platinum cathode of the Clark electrode.

Updike and Hicks [4] improved upon the Clark-Lyons-Sachs electrode by trapping glucose oxidase in the pores of a thin acrylamide gel, photo-polymerized *in situ*, over the plastic membrane of a Clark type oxygen electrode. By this technique the dialysis membrane and enzyme layer are combined in a single, thin gel-matrix layer. The shorter diffusion paths for oxygen and glucose in this sensor result in a faster response. The time required here to achieve 98 percent of the steady state response varies from about 30 seconds to 3 minutes depending primarily on the thickness of the gel and plastic membranes. Updike and Hicks [4] suggested that their device could be used for *in vivo* monitoring of tissue glucose concentration. However, major problems in adapting such a sensor for long-term *in vivo* operation include drift and instability of the Clark type polarographic electrode [5], instability of the glucose oxidase enzyme in the gel [6], and protein deposition and/or blood coagulation phenomena on the surface of the enzyme matrix.

As noted by Severinghaus [5], much of the instability of the Clark oxygen electrode is associated with gradual cold flow and loosening of its plastic membrane, which initially is tightly stretched over the cathode tip. With its membrane loose, the electrode responds slowly, shows increased current, and becomes sensitive to minor differences in hydrostatic pressure. Even in its initial state where the membrane is tight, slight pressures from body movements can alter the thickness of the electrolyte layer between membrane and cathode, causing large changes or oscillations of signal current.

To overcome some of these problems, Bessman and Schultz [7] developed the glucose sensor shown in Figure 1, consisting of two oxygen-sensitive silver-lead electrolytic cells built into a single plastic disk, covered by a polypropylene tape membrane. A smaller disk of a cloth-reinforced matrix containing covalently-bound cross-linked glucose oxidase is cemented to the membrane over one cathode. A similar matrix without enzyme is cemented over the other cathode to equalize the diffusion rates of oxygen in the absence of glucose. When the sensor is exposed to a body fluid, glucose diffuses into the enzyme matrix and reacts with dissolved oxygen to decrease the rate of oxygen diffusion to the underlying cathode. At any constant level of dissolved oxygen in the ambient fluid, the difference or ratio of signal currents from the two cells is a function of glucose concentration.

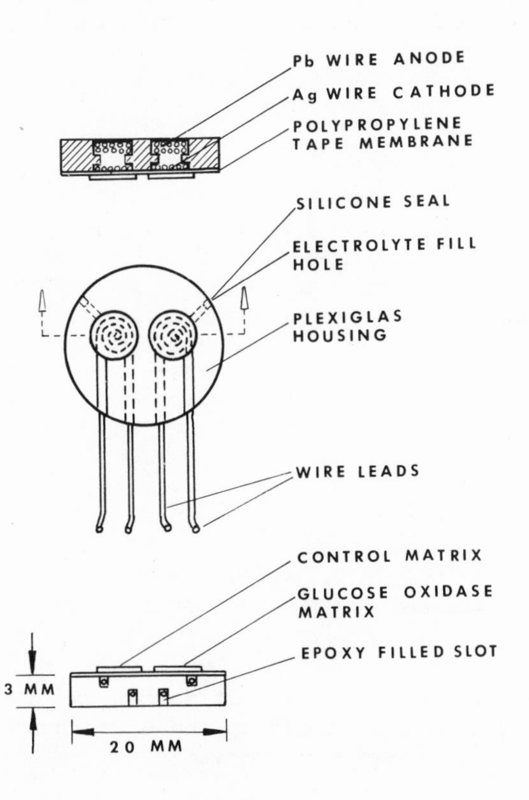

Figure 1: Bessman-Schultz enzyme-type dual cell glucose-oxygen sensor.

The operation of each oxygen-sensitive primary cell built into the Bessman-Schultz sensor is similar to that of the "self-generating" type of oxygen electrode described by Johnson, et al [8,9]. Each cathode is made of thin silver wire tightly wound into a flat spiral. Oxygen diffuses through the polypropylene tape membrane, then through the electrolyte between successive turns of the spiral and reaches the far and near surfaces of the cathode. Accordingly the mean effective thickness of the oxygen diffusion layer in the adjacent electrolyte is much larger than in the Clark electrode and therefore is less affected by small mechanical forces even in an aged sensor with a slightly loosened membrane. Figure 2 shows the relative effects of mechanical disturbance on such an aged sensor and a Yellow Springs Instrument Co. Type 4004 Clark electrode fitted with similar matrices. This aged Bessman-Schultz sensor shows temporary current increases or spikes that are only about 10 percent of the base signal, compared to about 54 percent for the Clark electrode, when stroked with the smooth rounded end of a plastic rod. The mechanical stability of a new Bessman-Schultz sensor is even more impressive because it is possible to run a finger nail over the matrix-membrane-cathode assembly without any change in signal.

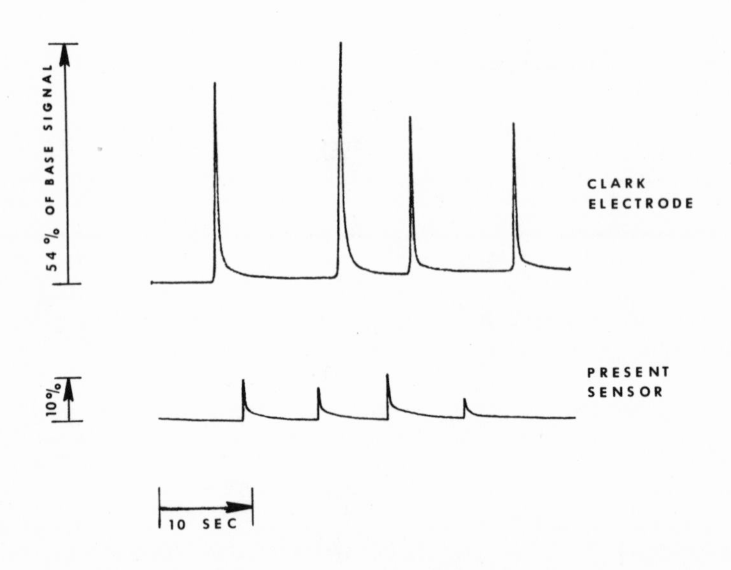

Figure 2: Comparison of the effect of mechanical disturbance on a Clark pO_2 electrode and on the Bessman-Schultz enzyme type glucose-oxygen sensor.

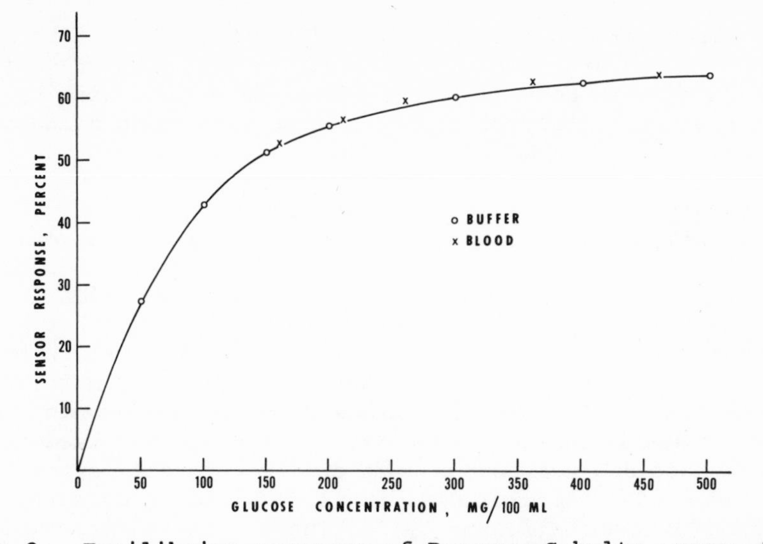

Figure 3: Equilibrium response of Bessman-Schultz enzyme-type sensor to glucose in flowing buffer solution and in flowing citrated bovine blood.

Figure 3 shows the response of the sensor to flowing buffer solution and citrated blood saturated with oxygen at 37°C. The percent response is defined as 100 [1-(I oxidase/I control)] where I oxidase is the current from the cell equipped with the enzyme matrix and I control the current from the cell with the enzyme-free matrix.

The response time of a Bessman-Schultz sensor with a polypropylene membrane and acrylic adhesives is about 6 - 7 minutes for sensing 90 percent of the change from 100 to 150 mg glucose/ 100 ml. With silicone membrane and adhesives the response time is shorter by about a half. In this case, each cathode area must be decreased considerably to avoid a current excess that can shorten the life of the sensor.

The enzyme matrix of the Bessman-Schultz sensor is made of thin closely-woven rayon acetate cloth (coated on one side with water-proof pressure-sensitive adhesive) impregnated with a glucose oxidase mixture, dried in air, and cross-linked with glutaraldehyde. The resultant covalently-bound enzyme is very resistant to degradation by body fluids, showing less than 10 percent decrease in its response to physiological concentrations of glucose after subcutaneous implantation in a rabbit for 11 days and only about 50 percent loss in activity after 6 months. *In vitro*, however, in 500 mg percent glucose solution in buffer at 37°C, the activity disappears in about 3 days. The resistance of the bound enzyme to degradation in vivo is believed to be the result of rapid destruction by body fluids of the hydrogen peroxide by-product of the enzyme-catalyzed reaction of glucose with dissolved oxygen to yield gluconic acid.

Noble Metal Catalyst and Fuel Cell Sensors

Prior to their development of their stabilized glucose oxidase matrix [7,10]. Bessman and Schultz [11] explored the use of noble metal alloy catalysts as substitutes for glucose oxidase in a sensor of sugar in body fluid. Earlier work had shown that a variety of sugars can be oxidized at room or body temperature by dissolved oxygen with the aid of platinum or other noble metal catalysts. D-glucose, for example, is oxidized to D-gluconic acid on a platinum-charcoal catalyst [12]. Similarly, D-galactose, D-xylose, D-mannose and L-arabinose are oxidized to the corresponding aldonic acids. Accordingly, it is possible to convert an oxygen probe to a sugar sensor by coating the oxygen-permeable membrane with an appropriate noble metal catalyst. Figure 4 shows a diagram of a Clark oxygen probe converted to a sugar sensor by a layer of 85 percent (atomic) Au-15 percent Pt alloy black catalyst secured to the membrane by a thin layer of an acrylic hydrogel adhesive. When immersed in a glucose-serum-buffer solution

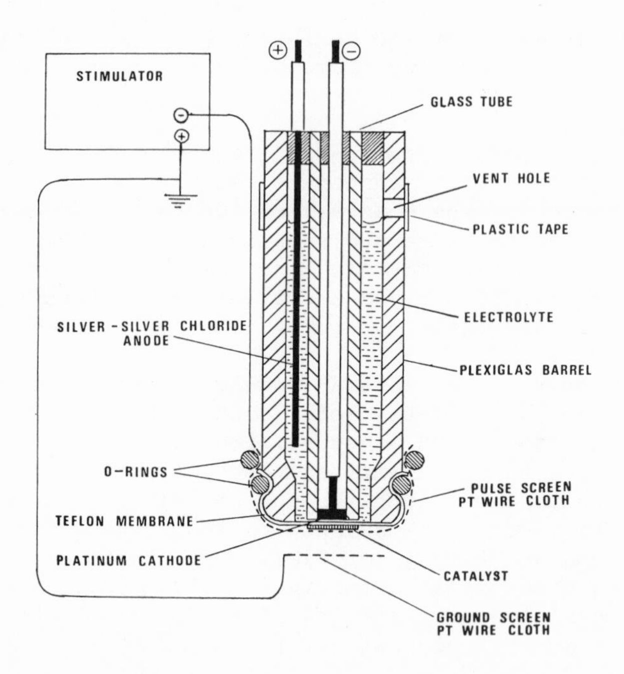

Figure 4: Bessman-Schultz catalyst-type sugar sensor.

approximating the composition of extra-cellular fluid, the catalysts become poisoned by adsorption of sulfhydryl, amino groups or oxide. In this condition, the sensor is completely unresponsive. However, application of negative voltage pulses to a platinum wire cloth strip in contact with the catalyst layer can regenerate the catalyst by reducing and removing the poisons from the active sites. However, repeated consistent regeneration of the catalyst is difficult because of fouling by precipitation of protein or other constituents of the body fluids.

A similar catalyst poisoning problem was encountered by Chang, et al [13] in their development of a sugar sensor based on earlier *in vivo* sugar-fuel cell designs by Wolfson, et al [14] and Drake, et al [15]. In the sugar-sensitive fuel cell the anode consists of either platinized wire mesh or a porous sheet of platinum black separated from the body fluid by a semi-permeable dialysis membrane. Unfortunately low molecular weight poisons like urea diffuse through the dialysis membrane and deactivate the anode. Accordingly an additional wire mesh "pulsing" electrode is placed against the body fluid side of the dialysis membrane. Application of positive and negative voltage pulses between the anode and the pulsing electrode serves to rejuvenate the catalyst. However, after each pulsing treatment, a recovery period of about 5 to 15 minutes is required for desorption of gluconic acid poison and for steady-state ionic and reactant gradients to be restablished in the fuel cell before a current indicative of sugar concentration can be attained [16]. It is not clear from the published description of Chang, et al [13], how many times and how consistently the catalyst in their sensor can be regenerated by electrical pulsing.

The Biocompatibility Problem

As noted by Chang, et al [13], fibrous tissue encapsulation is a serious problem for any glucose or sugar sensor implanted *in vivo* because it gradually chokes off the supply of both oxygen and glucose to the sensing surface. They suggest that membrane occlusion by fibrin deposition in an implant environment can be avoided if the external surface of the sensor is coated with a thin sulfonated polyelectrolyte-complex gel. However, much more work on this biocompatibility problem will be necessary to make the artificial pancreas a reality.

References

1. Clark, L.C., Jr. and Lyons, C. Electrode systems for continuous monitoring in cardiovascular surgery. Ann. N.Y. Acad. Sci., **102**:29, 1962.

2. Clark, L.C., Jr. and Sachs, G. Bioelectrodes for tissue metabolism. Ann. N.Y. Acad. Sci., 148:133, 1968.

3. Clark, L.C., Jr. Monitor and control of blood and tissue oxygen tensions. Trans. Am. Soc. Artificial Internal Organs, 2:41, 1956.

4. Updike, S.J. and Hicks, G.P. The enzyme electrode. Nature, 214:986, 1967.

5. Severinghaus, J.W. Measurements of blood gases: pO_2 and pCO_2. Ann. N.Y. Acad. Sci., 148:115, 1968.

6. Hicks, G.P. and Updike, S.J. The preparation and characterization of lyophilized polyacrylamide enzyme gels for chemical analysis. Anal. Chem., 38:726, 1966.

7. Bessman, S.P. and Schultz, R.D. Prototype glucose-oxygen sensor for the artificial pancreas. Trans. Am. Soc. Artificial Internal Organs, 19:361, 1973.

8. Johnson, M.J., Borkowski, J. and Engblom, C. Steam sterilizable probes for dissolved oxygen measurement. Biotech. Bioengr., 6:457, 1964.

9. Borkowski, J.D. and Johnson, M.J. Long-lived steam-sterilizable probes for dissolved oxygen measurement. Biotech. Bioengr., 9:635, 1967.

10. Bessman, S.P. and Schultz, R.D. Stabilized glucose oxidase electrode for monitoring glucose in biological fluids. Digest. Third Int. Congr. Med. Phys. Med. Engrg. (Eds. R. Kadefors, R.I. Magnusson and J. Petersen), Chalmers Univ., Göteborg, Sweden, 1972. Art. 30.6.

11. Bessman, S.P. and Schultz, R.D. Sugar electrode sensor for the "artificial pancreas". Horm. Metab. Res. 4:413, 1972.

12. Heyns, K. and Paulsen, H. Selective catalytic oxidations with noble metal catalysts. W. Foerst, Newer Methods of Preparative Organic Chemistry, 2:303, 1963.

13. Chang, K.W., Aisenberg, S., Soeldner, J.S. and Hiebert, J.M. Validation and bioengineering aspects of an implantable glucose sensor. Trans. Am. Soc. Artificial Internal Organs, 19:352, 1973.

14. Wolfson, S.K., Jr., Yao, S.J., Geisel, A. and Cash, H.R., Jr. A single electrolyte fuel cell utilizing permaselective membranes. Trans. Am. Soc. Artificial Internal Organs, 16: 193, 1970.

15. Drake, R.F., Kusserow, B.K., Messinger, S. and Matsuda, S. A tissue implantable fuel cell power supply. Trans. Am. Soc. Artificial Internal Organs, 16:199, 1970.

16. Personal conversations with J. Stuart Soeldner, April, 1972, and Kuo Wei Chang, August, 1973.

DISCUSSION

Question: (Hebert)

I understand that you have also developed a metal catalyst which alleviates the use of glucose oxidase. Would you describe this catalyst for us?

Answer: (Schultz)

This catalyst is described in a recent edition of Hormone and Metabolic Research 1973 (Spring) in a publication by Bessman and Schultz.

Question: (Wright)

1. How do you envision implanting the sensor in a diabetic individual?
2. What is the relative sensitivity of the sensor to oxygen versus glucose?

Answer: (Schultz)

1. Either in an artificial shunt in the blood stream or subcutaneously. However in either case there are biocompatibility problems in preventing fouling of the matrices by platelets and/or protein.

2. Overall the device is more sensitive to oxygen concentration than glucose. However in the normal physiological range 100-150 mg glucose/100 ml we find reasonably good sensitivity to glucose. When we drop down from 150mm pO_2 to 40mm pO_2, we lose the ability to distinguish between 200 and 250 mg glucose/100 ml, but we still retain sensitivity in the normal physiological range.

INDEX

ION-SELECTIVE MICROELECTRODES

Edited by Hebert J. Berman
Boston University
and Normand C. Hebert
Microelectrodes, Inc.

Volume 50 in
Advances in Experimental Medicine and Biology

Because of their unique capability for measuring chemical species without altering natural or controlled environmental conditions, pH- and ion-selective microelectrodes are finding numerous applications in the study and control of living systems. Providing insight into some of the techniques, present limitations, and applications of these tools, this work analyzes such topics as the fabrication of reference and glucose electrodes, the measurement of intercellular ion activity and calculation of equilibrium potentials, and techniques for the fabrication of pH, antimony, oxygen, and single- and double-barreled potassium and chloride liquid ion-exchanger microelectrodes.